SHEET-METAL PATTERN DRAFTING

AND

SHOP PROBLEMS

By

JAMES S. DAUGHERTY

COLLEGE OF INDUSTRIES
CARNEGIE INSTITUTE OF TECHNOLOGY

PITTSBURGH, PA.

THE MANUAL ARTS PRESS
PEORIA, ILLINOIS

Sheet-Metal Pattern Drafting
& Shop Problems

by James S. Daugherty

Originally published by
The Manual Arts Press, Peoria IL

Original Copyright 1922
by James S. Daugherty

Reprinted by Lindsay Publications Inc
Bradley IL 60915

ISBN 1-55918-294-6

2003

1   2   3   4   5   6   7   8   9   0

PREFACE

THIS book has been prepared as a text for use in vocational schools, trade schools, technical schools and high schools offering courses in sheet-metal pattern drafting and shop work, and for home study by apprentices and sheet-metal workers. It meets every requirement as a text book, and is also well adapted for reference use by draftsmen, shop foremen, and metal workers engaged in laying out patterns for general sheet-metal work, heating, ventilating, cornice, skylight, and heavy plate work.

The problems are practical and easily adaptable to varying courses in sheet-metal work, and have proved of exceptional value where pattern drafting and shop work are correlated and taught in a thoro, systematic manner in the best vocational and technical schools.

The subject-matter and method of presentation are the outcome of many years of teaching and practical experience in the various branches of the sheet-metal industry.

The proper sequence so necessary for successful instruction in sheet-metal pattern drafting is an important feature of this book; also, each problem presented is drawn to scale or dimensions given, and is of ample size for constructing from metal, using a minimum amount of material.

The descriptions are clear and well organized step by step. They stimulate the student to think and reason, and simplify the instructor's work. Numerous illustrative problems are worked thruout the text, and a large number of examples for the student are given in each chapter. The work is so planned that the student can take care of himself to a great extent.

The demonstrations of these practical problems are rendered clear by the illustrations which are photographic reproductions of work in which the patterns were developed, transferred to metal, and constructed by students in the sheet-metal department of the Carnegie Institute of Technology.

JAMES S. DAUGHERTY.

College of Industries,
Carnegie Institute of Technology.
November, 1921.

CONTENTS

A View of the Sheet-Metal Shop, Carnegie Institute of Technology.

PART ONE
Drafting Principles

CHAPTER I
Drawing Equipment

Equipment.—The following list comprises the equipment required for a course in sheet-metal pattern drafting: Drawing board, 24″ x 30″; T square, 30″; 45° triangle, 10″; 30° x 60° triangle, 10″; architect's scale, 12″; 4H drawing pencil; pencil erasing rubber; thumb tacks; detail paper; also a set of drawing instruments consisting of the following pieces: 5″ dividers, 5½″ compasses, 3″ bow spacers, 3″ bow pencil, ruling pen, bow pen and irregular curves.

Paper.—The paper in general use for sheet-metal pattern drafting is known as brown detail paper. It can be bought of almost any width, in large or small rolls, and is sold by the yard or pound. The paper should be of medium thickness, very strong and tough, because a shop drawing is likely to be subjected to considerable rough usage. If a finished drawing is to be made, white drawing paper should be used. It can be obtained in almost every conceivable grade and a variety of sizes.

Pencils.—For working drawings, full size details, etc., on manilla paper, a 4H pencil is quite satisfactory. For developing miter patterns in which the greatest accuracy is required, a 5H pencil is generally used. The accuracy of drawings depends, in a great measure, upon the manner in which the pencils are sharpened. To sharpen the pencil, remove the wood from both ends by means of a sharp knife, exposing about 3/8″ of lead. One end should then be sharpened to a round point, and the other to a chisel point or a wedge-shaped end. This operation should be done with a fine file or pencil sharpener. A strip of No. 0 sandpaper, 4″ long and 3/4″ wide, glued upon a thin strip of wood, will be found very serviceable.

The chisel end is used for drawing straight lines, and the conical point for free-hand sketching and marking dimensions. A soft pencil should never be used for drawing, because it becomes dull after drawing a few lines. This makes it impossible to draw fine, sharp lines and keep the paper clean.

Preparation of the Paper.—The paper is fastened to the board by means of thumb tacks, and care must be taken to have it lie perfectly flat on the board, so that it will have no wrinkles. To do this, proceed as follows: Place the long edge of the paper parallel with the long edge of the board, the paper being within about three inches of the lower and left-hand edge of the board. Insert a thumb tack in the upper right-hand corner and press it in until it is flush with the surface of the paper. Next, place the left hand on the paper near the upper right-hand corner; then slide the hand toward the lower left-hand corner, removing all wrinkles, and insert a thumb tack as before. Lay the left hand on the middle of the sheet and slide it toward the upper left-hand corner; holding it there, press in the third tack. Slide the hand from the center of the paper toward the lower right-hand corner, and insert the fourth tack, completing the operation. During damp weather, when conditions are such as to cause the paper to swell and become very wavy and loose, remove three of the tacks and again fasten the paper as before, squaring the sheet by the most important line on the drawing.

SHEET-METAL PROBLEMS CONSTRUCTED UNDER DIRECTION OF THE AUTHOR.

In presenting this subject to the student, no attempt has been made to give a complete course in geometry. Problems have been selected which will be of the greatest assistance to the pattern draftsman, and which are composed of examples that are used in every-day practice. They are arranged in logical order, and teachers and students will find this selection to cover the subject properly with reference to order and extent.

Practical geometry is the science of geometry adapted to practical purposes by omitting theoretical demonstrations. Every student whose aim it is to become a proficient sheet-metal draftsman should have a fair knowledge of the subject. The problems should be carefully studied and worked with great accuracy, as the technical skill acquired in the use of the drawing instruments will be of great value in later work.

Geometrical Problems

When the problems herein given have been carefully studied, draw each problem, completing each step in the construction before proceeding to the next. All lines should be as sharp and fine as consistent with clearness.

In the geometrical figures, the given and required lines are shown in full heavy lines, and the construction in full light lines.

Preparation of Plates.—The size of paper recommended for the problems of this course is 15″ x 20″. The size of each plate is to be 14″ x 18″, having a border line all around 1/2″ from the edge of the plate, leaving the space inside of the border line 13″ x 17″. Divide the plate into two equal parts by means of a horizontal line. Using the scale, divide the length of plate into three equal parts, as shown by the vertical lines. This divides the drawing plate into six rectangular spaces. The problems should be drawn as near the center of each space as possible.

Fig. 1. To bisect a straight line MN, or the arc of a circle MON.—Let MN 3-1/4 inches long be the given line which it is required to bisect. With centers M and N, and any radius greater than one-half of MN, describe the arcs *1* and *2*. Through the points of intersection of these arcs draw a line, and the points of intersection with the given line MN, and the arc MON, shown by OA will give the required points.

Fig. 2. To erect a perpendicular from a given line.—Draw the line AB about 3-1/4 inches long.. Locate point C near the middle of line AB. With the point of the compass on A and any radius greater than AB, describe an arc at *1*. On B, with the same radius, describe the arc *2*. Through the intersection of arcs *1* and *2*, draw the line EC, which will be the required perpendicular.

Fig. 3. To erect a perpendicular near the end of a given line.—Draw AB 3-3/4 inches long. About 1/2 inch from B locate the point C, from which the perpendicular is to be erected. With C as center and with any convenient radius, describe the arc *1-2*. Using the same radius, step off this distance from *1* to *3* and *3* to *4*. Using any radius with *3* and *4* as centers, describe arcs *5* and *6* intersecting each other at *7*. Draw a line from C thru *7*, which will give the required perpendicular at the given point C.

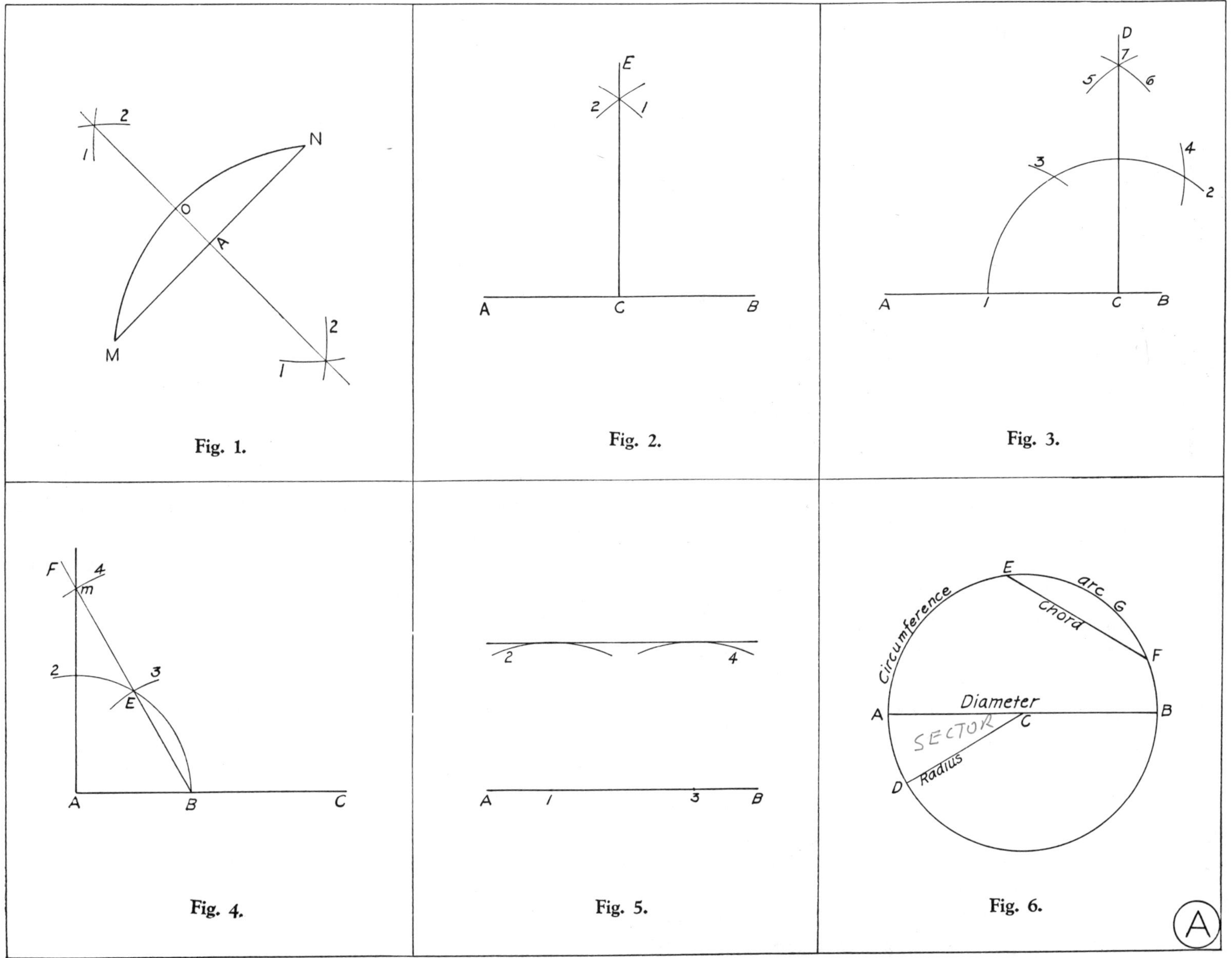

N
2
1
O
A
M
1
2
Fig. 1.
E
2
1
A
C
B
Fig. 2.
D
7
5
6
3
4
2
A
C
B
Fig. 3.
F
4
m
2
3
E
A
B
C
Fig. 4.
2
4
A
3
B
Fig. 5.
Circumference
E
arc
Chord
F
Diameter
A
B
SECTOR
C
Radius
D
Fig. 6.

Fig. 4. To erect a perpendicular at the end of a given line.—Draw *AC*, the given line, 3-3/4 inches long. Set the point of compass on *A*, and with any radius describe the arc *B2*. On *B* with the radius *AB*, describe the arc *3* intersecting arc *B2* at *E*. Thru *E* draw line *BF* indefinitely; with radius *BE*, describe arc *4*, intersecting line *BF* at *M*. Connect *MA*, which will be the perpendicular required.

Fig. 5. To draw a line parallel to a given line.—Draw *AB* 3-1/2 inches long. Near the end of the line at *1*, set the point of the compass, and with a 2-inch radius, describe arc *2*. With the same radius on the point *3*, describe the arc *4*. Then a line drawn touching the arcs *2* and *4* will be parallel to *AB*.

Fig. 6. To draw a circle and its properties.—Draw *AB* 3-3/4 inches long. Bisect *AB* at *C*. With point of compass at *C*, and radius *CA*, describe the circumference of the circle. The diameter of a circle is any straight line drawn thru the center to opposite points of the circumference as *AB*. The radius of a circle is any line as *CA* and *DC*, drawn from the center to any point in the circumference; two or more such lines are radii, the plural of radius. An arc of a circle is any part of the circumference as *EG*. The sector of a circle is the part of a circle included between the radii and the arc which they intercept, as *ACD*. A segment of a circle is a part cut off by a chord, as *EFG*. A chord of a circle is a straight line joining the extremeties of an arc, but not passing thru the center, as *EF*.

Fig. 7. To draw a tangent from any given point on a circle.—With point of compass at *A* and radius *AB*, describe a circle 3-1/2 inches in diameter. Thru point *B* and center *A* draw a straight line. A perpendicular drawn thru point *B* will give the required tangent, as *CD*.

Fig. 8. To bisect a given angle.—Draw the given angle *ABC*. With any convenient radius and *B* as center, describe the arcs *1* and *2*. With the same or a larger radius and *1* and *2* as centers, describe arcs intersecting at *3*. Draw a line from *3* to *B*, which divides the angle *ABC* into two equal parts. This problem shows how to obtain the miter line between the two parts of an elbow or sheet-metal molding.

Fig. 9. To draw an equilateral triangle, one side being given.—Draw *AB* 2-3/4 inches long. With *A* as center and *AB* as radius, describe arc *1*. With *B* as center and the same radius, describe arc *2* intersecting the former arc at *C*. Draw the lines *BC* and *AC*, and *ABC* is the required equilateral triangle.

Fig. 10. To construct an angle similar to a given angle.—Let *ABC* be the given angle. With *A* as center and with any radius, describe arc *1–2*, touching both sides of the angle. Draw line *EF* equal to *AB*. With *E* as a center and radius *A2* of the given angle, describe arc *3–4*. With *4* as center and radius *1–2*, describe arc *5* intersecting arc *3–4* at *G*. A line drawn from *E* thru point *G*, completes the angle equal to *ABC*.

Fig. 11. To draw a triangle equal to any given triangle.—Draw the given triangle *ABC* and line *1–2* equal to *AB*. With the radius *AC* and the center at *1*, describe the arc *3*. With the center at *2* and the radius *BC*, describe arc *4* intersecting arc *3* at *5*. Draw lines *1–5* and *2–5*, which will give the required triangle equal to the given triangle *ABC*.

Fig. 12. To construct an irregular angular figure similar to a given figure.—Draw line *AB* 3-1/4 inches long and construct trapezium *ABCD*. To copy this figure in exactly the same size as it is given, draw line *1–2* equal in length to *AB*. With *A* as center and *Ae* as radius, describe arc *ef*. With the same radius and *1* as center, describe arc *3–4*. With *e* as center, describe the arc *g*. With the same radius and *3* as center, describe arc *5* intersecting arc *4* at *6*. Draw a line from *1* thru *6*,

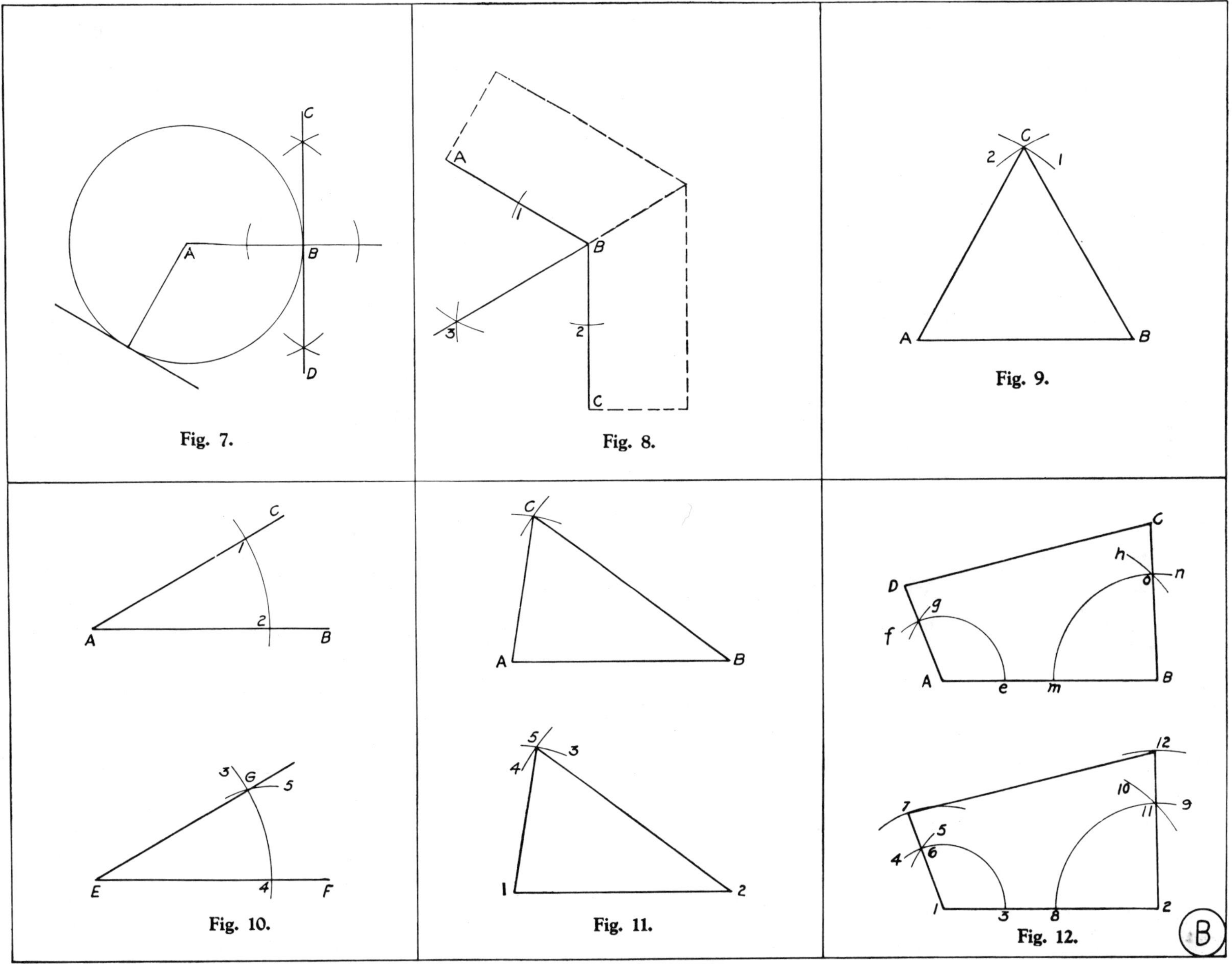
C
A
B
D
Fig. 7.
A
B
C
3
2
Fig. 8.
C
2 1
A
B
Fig. 9.
C
1
2
A
B
Fig. 10.
3 G 5
E
4
F
C
A
B
Fig. 11.
5 3
4
1
2
D
h
n
9
o
f
g
A
e
m
B
C
Fig. 12.
12
10
9
7
11
5
4 6
1
3 8
2
B

making *1–7* equal to *AD*. With *B* as center, describe arc *mn*. With the same radius and *2* as center, describe the arc *8–9*. With *m* as center, describe arc *h*, cutting line *BC* at *o*. With the same radius and *8* as center, describe arc *10* intersecting arc *9* at *11*. Draw line *2–12* thru point *11*, and make it equal in length to *BC*. Draw line *7–12* to complete the trapezium similar to the given trapezium *ABCD*.

Fig. 13. To construct a square from a given side.— Draw line *AB* 3-1/2 inches, the length of the given side. With *A* as center and *AB* as radius, describe arc *B–1* indefinitely. With the same radius and *B* as center, describe the arc *A–2*, intersecting arc *1* at *C*. Bisect *AC* at *D* thru intersecting arcs at *E*. With *C* as center and radius *CD*, describe arcs *3* and *4*, intersecting arcs *1* and *2* at *F* and *G*. To complete the square, connect *AF*, *FG* and *GB*.

Fig. 14. To construct a regular pentagon in a given circle.—With *A* as center and with the compasses set to 1-7/8 inches, describe the circle *BCDE*. Draw the two diameters *BD* and *EC* perpendicular to each other. Bisect the radius *AB* by the line passing thru *AB* at *1*. With *1* as center and *1–E* as radius, describe the arc, locating point *2*. With *E* as center and the distance *E–2* as radius, describe an arc cutting the circumference of the circle at *3* and *4*. Using the same radius with *3* and *4* as centers, describe the arcs *5* and *6*. Connect points *E–4–5–6–3*, which completes the pentagon.

Fig. 15. To construct a pentagon from a given side.— Let *AB* be the given side. With *A* as center and with the compasses set to 1-3/4 inches, describe the semi-circle *BEC*. Divide *BEC* into five equal parts, and from *A* draw lines thru the divisions *1–2–E*. With *AB* as radius and *E* as center, describe the arc *3*. With the same radius and *B* as center, describe the arc *4*. Draw the lines *E–3*, *3–4* and *4–B* to complete the figure.

Fig. 16. To construct a hexagon from a given side.— Describe a circle with the radius *AB* 1-7/8 inches, which will be the length of the given side. Draw the diameter *BC*. With the radius *AB* and the centers *C–B*, describe the arcs *1–2–3–4*. Connect by straight lines *C–1*, *1–2*, *2–B*, *B–3*, *3–4* and *4–C*, which completes the required hexagon.

Fig. 17. To inscribe an octagon within a given circle.— With *A* as center, with the compasses set to 1-7/8 inches, describe the circle *1–2–3–4–5–6–7–8*. Let this be the given circle in which it is required to inscribe a regular octagon. Thru the center, draw lines *BC* and *DE* perpendicular to each other, cutting the circumference of the circle at *1–5* and *7–3*. Bisect the angles *DAB* and *DAC*, and let the bisector of each angle meet the circumference at *2* and *8*. Draw the diameters *8–4* and *2–6*. Straight lines drawn from *1–2*, *2–3*, *3–4*, etc., will form the required octagon.

Fig. 18. To inscribe an octagon within a given square. —Draw line *AB* 4 inches long, and construct the given square *ABCD*. Draw diagonal lines *DB* and *AC*. With *B* as center and *BG* as radius, describe the arc *1–2*. With the same radius and *ADC* as centers, describe arcs *3–4*, *5–6*, *7–8*. Straight lines drawn from *6–2*, *2–8*, *8–3*, etc., will complete the required octagon.

Fig. 19. To construct an octagon, one side being given. —Draw line *AB* 1-1/4 inches long, which is the length of the given side. Extend *AB* indefinitely, as shown by *1* and *2*. From *A* and *B* erect indefinite perpendiculars as *AC* and *BD*. With *A* and *B* as centers, using any radius, draw the arcs *1–3* and *4–2*. Bisect the angles *1–A–3* and *4–B–2* by *5–A* and *B–6*. On these two lines set off *A–7* and *B–12*, equal to *AB*. From *7* and *12* erect the perpendiculars *7–8* and *12–11*, equal to *AB*. With *8* and *11* as centers and *AB* as radius, describe arcs *9–10*, inter-

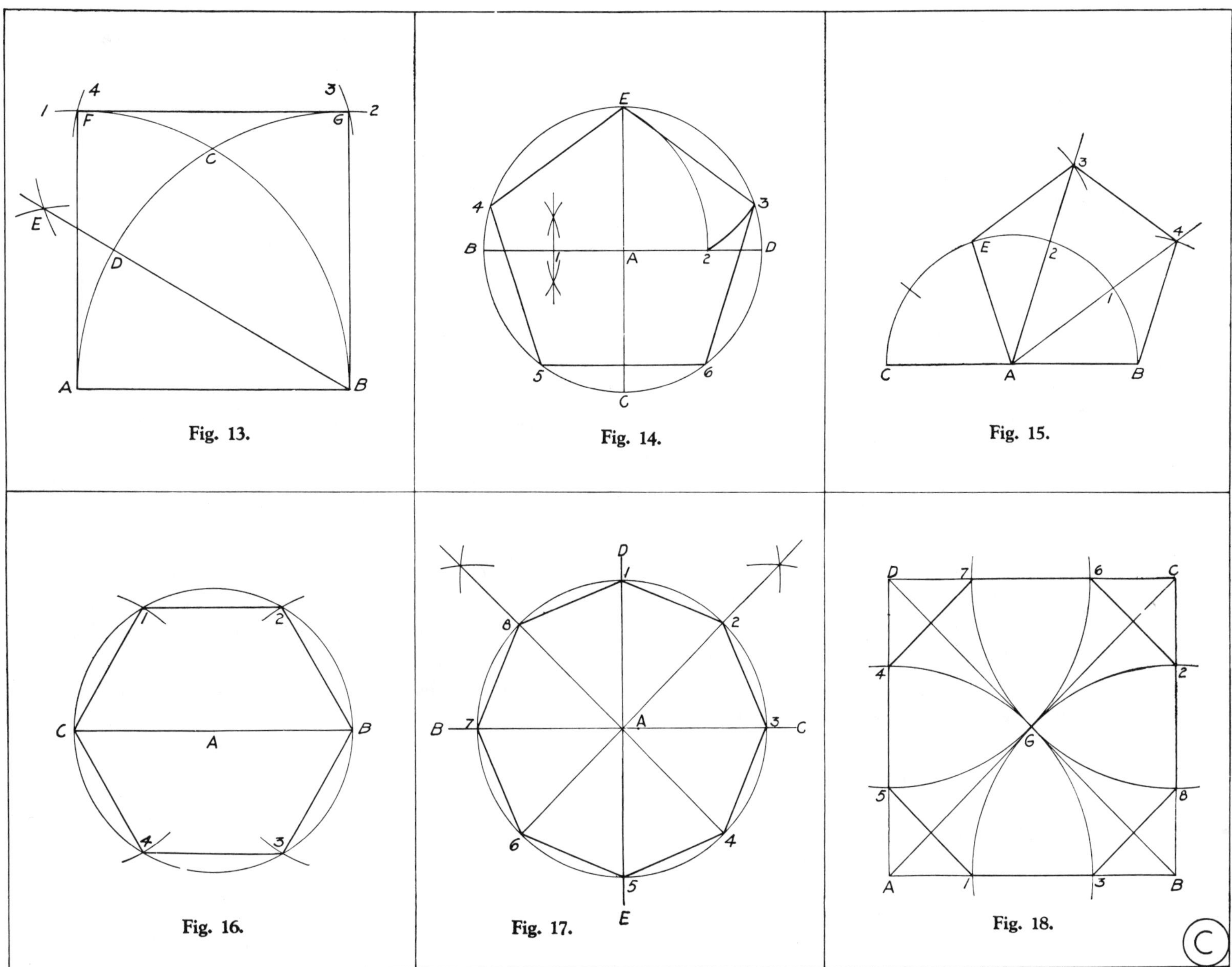

Fig. 13.
Fig. 14.
Fig. 15.
Fig. 16.
Fig. 17.
Fig. 18.

secting perpendiculars *AC* and *BD* at *9* and *10*. Connect *8–9*, *9–10* and *10–11*, which completes the required octagon.

Fig. 20. To draw a circle thru any three given points not in a straight line.—Let *ABC* be the given points not in a straight line. Draw the lines *CA* and *AB*. Bisect the line *CA* by *EF*, as shown. Also bisect *AB* by the line *GF*, and the intersection of the bisecting lines at *F* will be the center of the required circle. Then with *F* as center and *FB* as radius, describe the circumference thru points *ABC*.

Fig. 21. To find the center of a circle when the circumference is given.—Let *ABC* be the given circle. From any point on the circle as *B*, with any radius, describe the arc *1–2*. Then from the points *A* and *C*, with the same radius, describe the intersecting arcs *3–4* and *5–6*. Thru the points of intersection draw the lines *7–8* and *9–10*, which will meet in *x*. Then *x* will be the center of the circle.

Fig. 22. To describe the segment of a circle of any given chord and height.—Draw the line *AB* 3-3/4 inches long, which will be the given chord. Draw the perpendicular *mn* indefinitely, and make *Pm* the given height 1-1/8 inches long. Connect *mB* and bisect *mB* by the line *CG*, intersecting the perpendicular *mn* at *C*. Then *C* will be the center from which to describe the segment *AmB*.

Fig. 23. To inscribe an equilateral triangle in a circle.—Draw the line *AB* 3-3/4 inches long, which will be the diameter of the given circle. With *B* as center and *BC* the radius of the circle as radius, describe the arc *FCG*. To complete the inscribed triangle, connect by straight lines *FA*, *AG* and *GF*.

Fig. 24. To inscribe a circle in a given triangle.—Draw the line *AB* 3-3/4 inches long. Make *AC* 3 inches, and *CB* 4 inches in length. Bisect the angles *CAB* and *ACB*. The intersection of the bisectors at *m* will be the center of the circle which can be described, touching all three sides of the triangle. The sides *AB*, *BC* and *CA* will be tangent to this circle.

Fig. 25. To draw an ellipse when the diameters are given, without using centers.—Draw the line *1–A* 3-1/2 inches long, which will be the required length. Bisect *1–A* at *C*. Thru *C* draw *DE* 2-1/4 inches long, the required width. With *C* as center and *C–1* and *CE* as radii, describe the outer and inner circles, respectively, as shown. Divide one-quarter of the outer circle into any convenient number of parts; in this case, into five, as shown by *1–2–3–4–5–6*. Divide the one-quarter inner circle into the same number, as shown from *1* to *6*. From the points on the smaller circle, draw horizontal lines, and thru the points on the larger circles, draw vertical lines. The points *a*, *b*, *c*, *d*, where the horizontal and vertical lines intersect, are points on the ellipse. Using an irregular curve, trace a line thru the points thus obtained, completing one-quarter of the ellipse.

Fig. 26. To draw an approximate ellipse when length and width are given, using circular arcs.—Draw the line *AB* 3-1/4 inches long. Bisect *AB* at *m*, and draw the width *CD* 2-1/8 inches long. On the length *AB*, set off the width *CD* from *B* to *3*, and divide the balance *3A* into three equal parts, as shown by *1*, *2*, *3*. With *m* as center and a radius equal to the length of two of these parts, describe arcs cutting *AB* in *E* and *F*. With *EF* as radius and *E* and *F* as centers, intersect arcs at *4* and *5*. Draw lines from *4* thru *E* and *F* as *4–6* and *4–7*, and lines from *5* thru *EF* as *5–8* and *5–9*. With *4* and *5* as centers and *5–C* and *4–D* as radii, describe the arcs *GDH* and *OCP*. With *E* and *F* as centers and radii equal to *EA* and *FB*, describe the arcs *GAO* and *PBH*, completing the ellipse.

Fig. 27. To draw an approximate ellipse, the major and minor axes being given.—For many purposes in sheet-metal drawing, it is sufficiently accurate to describe the ellipse

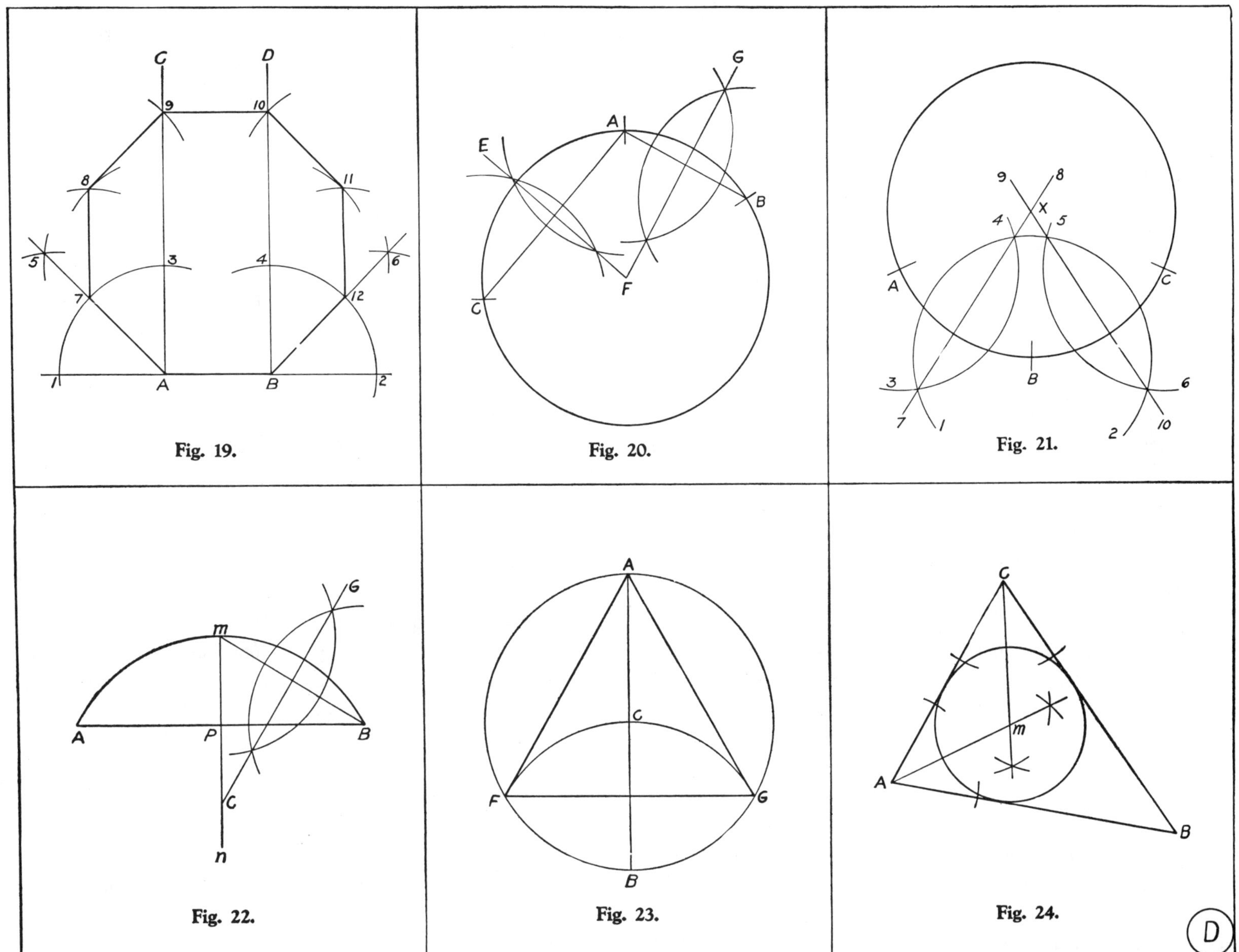

Fig. 19.

Fig. 20.

Fig. 21.

Fig. 22.

Fig. 23.

Fig. 24.

by means of circular arcs, and where centers must be used in developing patterns for flaring articles.   Draw the major diameter AB 3-7/8 inches long, and the minor diameter CD 3 inches in length.   On the line CD lay off mF and mG, equal to the difference between the major and minor diameters.   On the line AB lay off mE, and mH equal to three-quarters of mG.   Connect points $FHGE$, and extend the lines.   With center E and radius EA, describe arc RAO.   With center F and radius FD, describe arc ODP.   In a similar manner, describe arcs PBS and SCR from centers G and H.   This is not a practical method when the major diameter is more than twice the minor.

Fig. 28.   To draw an ellipse by intersection of lines.— Draw the major axis AB 3-1/2 inches long, and the minor axis mA' 2-1/4 inches.   Thru m parallel to line AB draw line CD. From points A and B erect perpendiculars to line CD.   Divide lines AC and DB into any number of equal parts; in this case, four, and draw lines from points 1, 2, 3, etc., to m.   Divide An and nB into the same number of equal parts, and draw lines from A' thru these points intersecting the similarly numbered lines drawn from the points on the line CA and DB.   Thru these points of intersection, trace the semi-ellipse AMB.

Fig. 29.   To draw an egg-shaped oval with arcs of circles.—With a radius of 1-3/8 inches and C as center, describe the circle $AmBn$.   Thru the center C, perpendicular to AB, draw the line mn.   Thru n draw Bn and An indefinitely.   On A and B as centers, with AB as radius, describe the arcs BH and AG.   With n as center, describe the arc GH to complete the figure.

Fig. 30.   To draw an ellipse by means of a pencil and thread.—Draw AB, the major axis, 3-3/4 inches long.   Bisect AB at m.   Thru m draw the perpendicular CD 2-1/2 inches long.   Take Am one-half the length of the major axis for radius,

and with C as center, describe the arc GH.   Drive pins at C, G and H; then tightly tie a thread around the three pins CGH. Remove the pin at C, and, placing a pencil at this point, keeping the thread tightly stretched, describe the ellipse.

Fig. 31.   To draw a parabola, having given the axis AB and the double ordinate FD.—Draw AB 3-1/2 inches long, and FD perpendicular to AB, 4 inches long.   Draw EF and CD parallel and equal to AB.   Divide EF and BF into the same number of equal parts.   From the divisions on BF, draw lines parallel to the axis AB, and from the divisions on EF, draw lines to the vertex A.   The points of intersection of these lines 1 and 1, 2 and 2, etc., are points on the required curve thru which it may be traced.   In like manner, obtain the opposite side.

Fig. 32.   To draw a hyperbola, the axis, a double ordinate and its distance from the vertex being given.—Draw the double ordinate FD 3-3/4 inches long.   Perpendicular to FD, draw the axis BA 3-3/8 inches long.   On the line AB locate M 1-3/8 inches from the vertex A.   Thru M draw EC perpendicular to AB; then draw FE and DC perpendicular to FD, intersecting FE and DC in E and C.   Divide FE and DC into the same number of equal parts, and from points 1, 2, 3, etc., on FE, draw lines to the vertex M.   From points on FB, draw lines to the vertex A.   The intersection of these lines 1 and 1, 2 and 2, etc., will be points in the required hyperbola.

Fig. 33.   To draw an equable spiral.—Draw the line $A6$ 4-5/8 inches long.   Bisect $A6$ at O, and with O as center and OA as radius, describe the circle $A3$–6–9. Divide the circle into twelve equal parts, and to the points on the circumference draw radial lines from the center at O.   Divide AO into as many equal parts as the spiral is to have revolutions; in this case, two.   Divide each space into twelve equal parts, the same number of parts as there are divisions in the circle.   With O as center and O–1, O–2

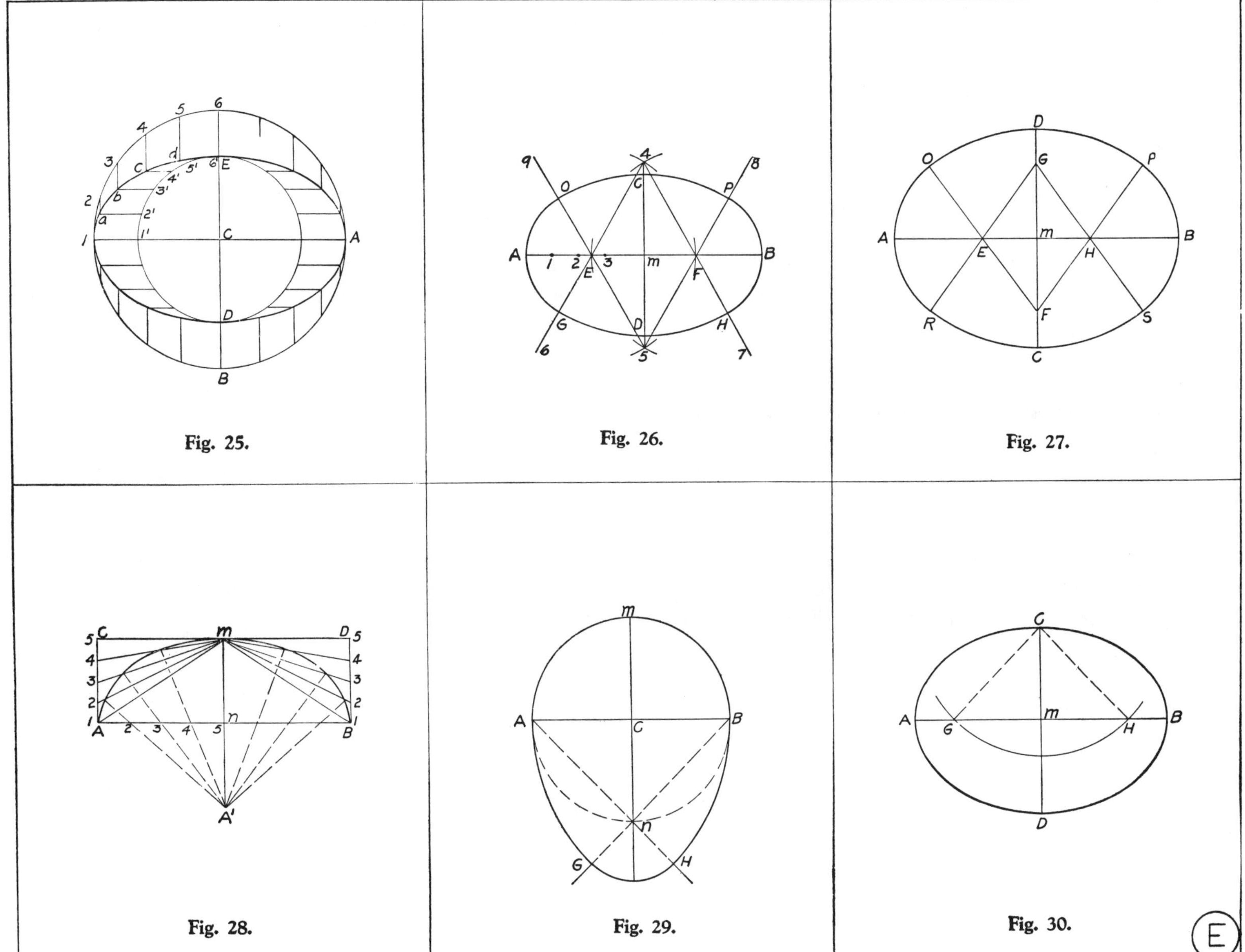

Fig. 25.

Fig. 26.

Fig. 27.

Fig. 28.

Fig. 29.

Fig. 30.

O–3, O–4, etc.,as radii, describe arcs intersecting similarly numbered radial lines, as shown. Thru the points of intersection thus obtained, trace a curved line, completing the required spiral.

Fig. 34. To draw a scroll to a specified width.—This problem shows how to obtain the pattern for a simple scroll much used in the construction of brackets and modillion blocks; also as a terminal finish in sheet-metal cornice construction. Draw *AB* the given width, 4-1/4 inches long. On the line *AB* locate point *C* one inch from *A*, for the width of the band. Make *AD* equal to one-third of *AB*. Bisect *AB*, obtaining the point *m*. Now, take one-eighth of the distance *AD* and set it off below point *m*, thus obtaining the point *G*. From *G* draw a horizontal line, and using the 45° triangle, draw from *A* a line intersecting the horizontal line at *F*. Draw the rectangle *GFEB*. From *E* draw a line, making an angle of 45° with *BE*, intersecting *GF* at *H*. Draw the square *GHOP*, and divide *GP* into four equal parts; on two of these parts, construct the inner square *1–2–3–4*. The arcs forming the outline of the scroll are described in the following manner: With *G* as center and radii *GC* and *GA*, describe the arcs *AF* and *C–6*. From *H* as center,

with radius *H–6* and *HF*, describe the arcs *FB* and *6–7*. Continue the operation by describing arcs tangent to those first drawn, using points *O*, *P*, and *1,2,3,4* at the angles of the smaller square as centers for describing the successive quadrants.

Fig. 35. The Helix.—The Helix is a curve formed by a point moving around a cylinder and at the same time advancing along the line of its axis a fixed distance for each revolution. The distance advanced at one revolution is called the pitch. The line described upon the surface of the cylinder is represented by a flexible cord wound around the cylinder; it is shown in actual practice by the thread of a screw. Let the circle *1–5–9–13*, having a diameter of 2-3/4 inches, represent a plan view of the cylinder. Draw the elevation *ABC–1*, and make *C–1*, the pitch of the helix, 3-1/2 inches long. Divide *C–1* into a number of equal parts; in this case, sixteen, and divide the circle into the same number of equal parts, beginning at point *1*, as shown in the drawing. From the points on the circle, as *1, 2, 3, 4*, etc., draw vertical lines, intersecting like numbered horizontal lines drawn from similarly numbered points on the pitch line *1–C*. Thru the points thus obtained, trace the helical curve, **making one revolution.**

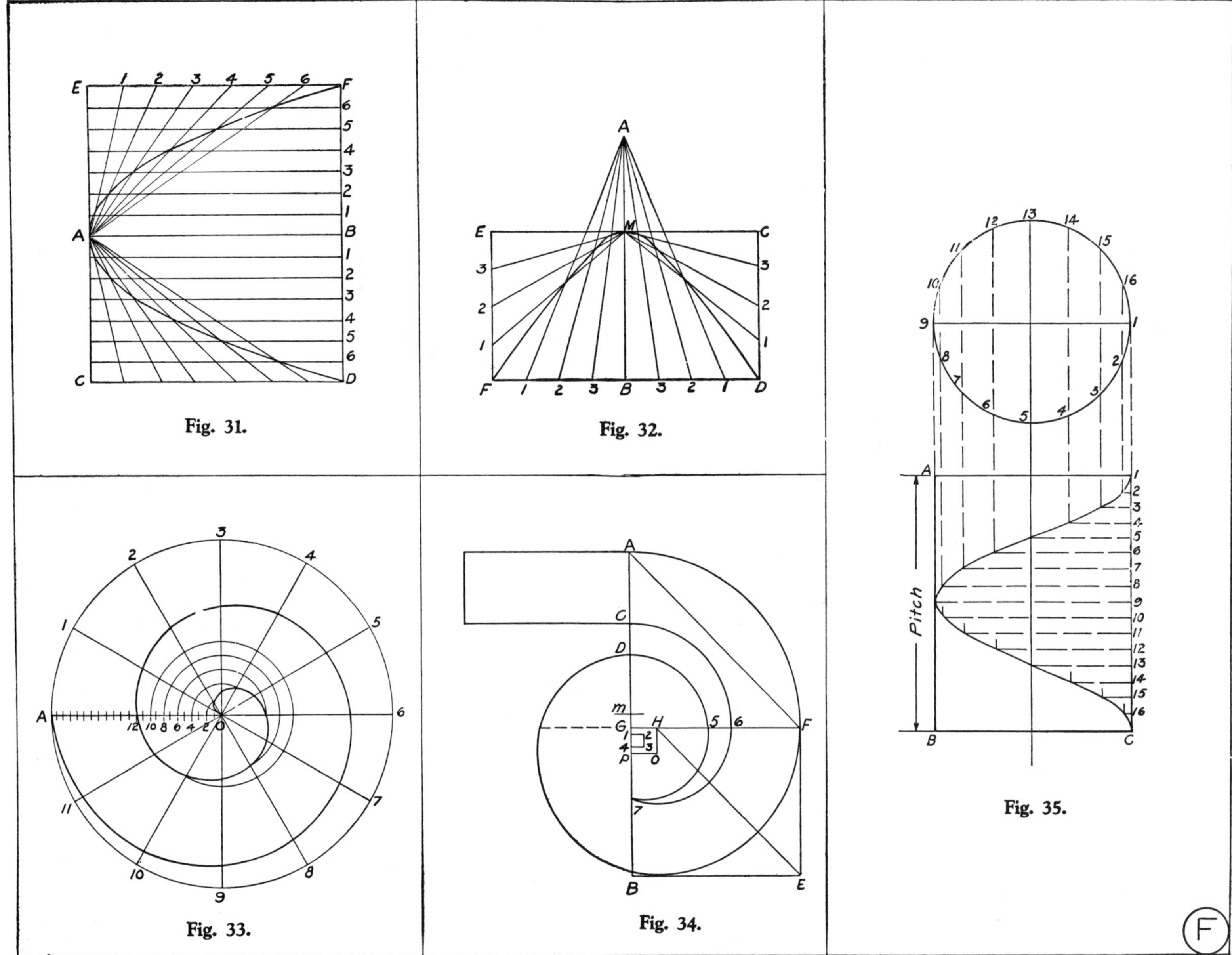

Fig. 31.
Fig. 32.
Fig. 33.
Fig. 34.
Fig. 35.

CHAPTER III
PRACTICAL PATTERN DRAFTING

Sheet-metal pattern drafting is founded upon those principles of geometry which relate to the surfaces of solids, and may be described as the development of surfaces. Sheet-metal articles are hollow, and are considered in the process of pattern drafting as though they were the coverings of solids of the same shape.

The different methods for developing the patterns for forms with which the pattern draftsman has to deal may be divided for convenience of description, into three general divisions:

First, *Parallel Line Developments*, which is used in developing patterns for moldings, pipes, elbows and regular continuous forms, or may be called parallel forms.

Second, *Radial Line Developments*. This method is used in developing patterns for regular tapering forms by means of radial lines, converging to a common center. The forms having for their base the circle, or any of the regular geometric figures which terminate in an apex.

Third, *Triangulation*. This method is used in developing patterns for irregular forms which cannot be developed by either the parallel or radial-line methods.

All of the problems that will follow should be carefully studied, drawn on detail paper, and the drawings proved by paper models when it is not possible to construct the problems from sheet metal. These models will at once show any error in the drawings which might otherwise be overlooked.

Practical work-shop problems, such as arise in every-day practice, are presented; an actual trade object forms the subject of each problem. No work is introduced that is not practical and likely to arise in the every-day work of a sheet-metal pattern draftsman. The problems should be taken in regular order, and drawings constructed as called for by the text.

Some of the problems are fully developed, and the demonstrations are made more explicit than others. Less important details of the work are sometimes omitted and certain parts of the operation are described in a general way upon the supposition that the problems in which the development is fully described would naturally be studied first. The problems should be drawn according to the dimensions given, and all problems will then be of ample size to permit each step in the drawing of the pattern to be clearly shown, and of proper size for constructing from sheet metal. Allowances for seams, joints, etc., are shown in some of the problems, but if a joint is required at a place other than where shown, it can be changed at the discretion of the draftsman without changing the principles involved in the development.

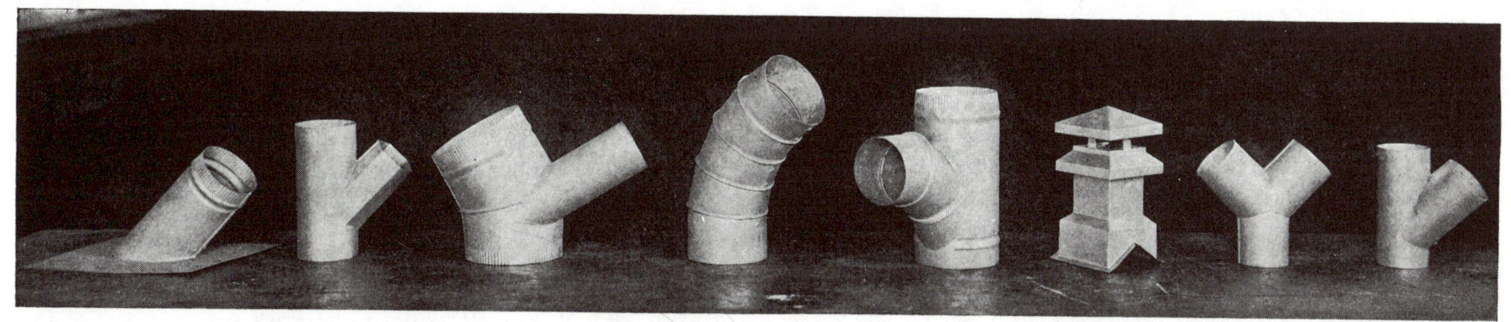

PART TWO

PARALLEL LINE DEVELOPMENTS

———

CHAPTER IV

PATTERN PROBLEMS

Under this head, problems are presented in which the patterns are developed by means of parallel lines. This method is used in developing the pattern for any form the opposite lines of which are parallel, such as elbows, T-joints, roof gutters, cornices, skylights, etc.; also in patterns for miters that occur in joining moldings, pipes and all regular continuous forms at any angle and against any other form or surface.

There are certain fixed principles that apply to developments by this method, and the following rules should be carefully observed by the student and draftsman.

1) A plan and elevation must first be drawn, showing the article in a right position, in which the parallel lines of the solid are shown in their true length.

2) The pattern is always obtained from a right view of the article in which the miter line or line of intersection is shown.

3) A stretchout, or girth, line is always drawn at right angles to the parallel lines of the solid, upon which is placed each space contained in the section or plan view.

4) Measuring lines are always drawn at right angles to the stretchout line of the pattern.

5) Lines drawn from the points of intersection on the miter line in the right view, intersecting similarly numbered measuring lines drawn from the stretchout, will give points showing the outline of the development.

6) A line traced thru the points thus obtained will give the desired pattern.

Problem 1. Pattern for Two-Piece Elbow in Round Pipe

In Figure 36, let *AB–1–CD–7* be the elevation of a two-piece 90° elbow. First, draw the elevation. Then, below the elevation, describe a circle representing the profile or plan, shown at *F*. As each half of the pattern is symmetrical, draw a line thru the plan *F*, and divide the upper half of the circle into a number of equal parts, as shown from *1* to *7*. From these points perpendicular lines are drawn, in-

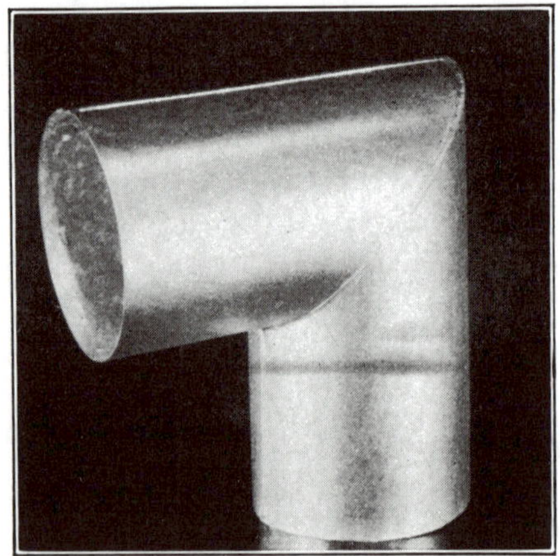

Problem 1.   Two-Piece Elbow, Round Pipe.

tersecting the miter line *1–7* as indicated. Then, at right angles to the vertical arm of the elbow *D–7*, draw the stretchout line *FG*, and upon this line step off twice the number of spaces shown in the plan, which will give the circumference of the elbow. From these points and at a right angle to *FG*, draw measuring lines which are intersected by like numbered lines drawn at a right angle to the cylinder from similarly numbered points on the miter line *1–7* in the elevation.   A line traced thru points

thus obtained will be the pattern for the vertical arm of the elbow, as shown by *FGLHJ*.

The irregular curve traced thru the points of the pattern is the only one required for both pieces of the elbow, and to save material, the pattern for the upper arm of the elbow is generally obtained in the following manner:

The stretchout of both pieces being of equal length, extend the outer lines of the pattern to *M* and *R*, as shown in the drawing, and make *JM* and *RL* equal in length to *A–7* in the elevation. Draw a line from *M* to *R*; then *JLRM* will be the pattern for the upper arm of the elbow, having the seam at *A–7* on the outside or heel of the elbow.   The seam on the lower arm is on the inner side or throat, as shown by *C–1* in the elevation.

This method of development is applicable to any pieced elbow, no matter what the shape of the pipe may be or the angle required.

Problem 2.   Conductor Elbow

In Figure 37, let *ABFRG* be the elevation of a two-pieced elbow, the circle representing the profile or plan being three inches in diameter.   Draw the elevation and make *BEF* an angle of 120°.   Bisect the angle *BEF* and draw the miter line *EG*. Divide the upper half of the profile into a number of equal parts, and from these points draw vertical lines intersecting the miter line *EG* in the elevation.

The stretchout line as shown by the line *ab* is next drawn at right angles to the lower arm of the elbow, and the patterns for both pieces are developed in the same manner as the 90° elbow, which is fully explained in Problem 1.

Problem 3.   Rectangular Conductor Shoe

The principle here involved and the method of procedure are exactly the same as in Problem 2, the only difference being in the

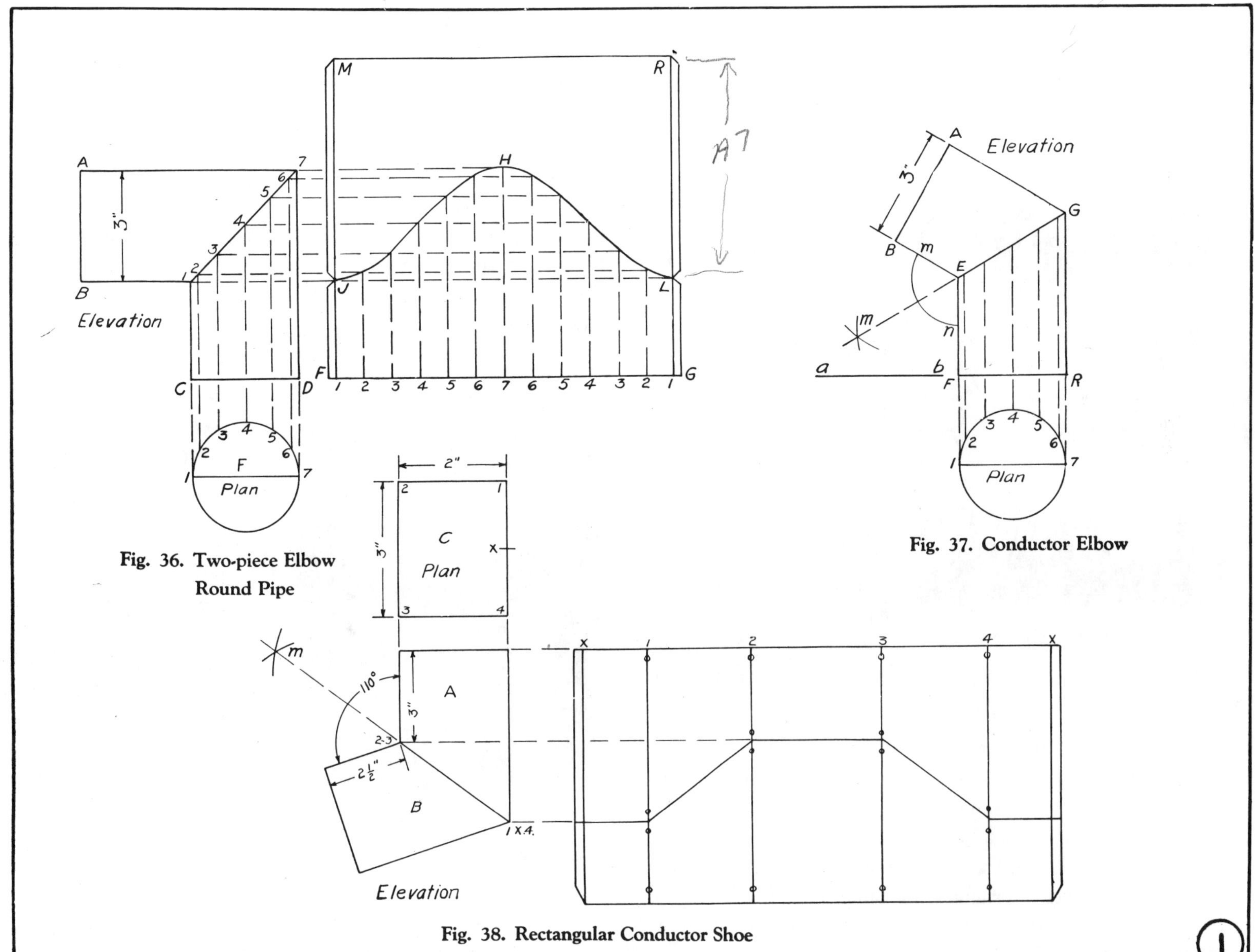

Fig. 36. Two-piece Elbow
Round Pipe

Fig. 37. Conductor Elbow

Fig. 38. Rectangular Conductor Shoe

shape of the pipe. In Figure 38, let *1–2–3–4–x* in the plan represent a rectangular pipe having the seam at *x*. Draw the plan and elevation according to the dimensions given on the drawing, and find the miter line by bisecting the angle in the usual manner.

Draw the stretchout line *x–x* at right angles to the vertical arm *A*, and on this line step off the spaces *x–1*, *1–2*, *2–3*, *3–4* and *4–x* of the plan *C*. From the points thus obtained draw the usual measuring lines, which are intersected by horizontal lines drawn from points *2–3* and *1–x–4* on the miter line in elevation. Lines drawn connecting these points will give the desired pattern.

The small circles marked near the extremity of each measuring line on the pattern indicate that the metal is to be bent along this line when forming the work into the required shape.

Problem 3.   Rectangular Conductor Shoe.

Problem 4.   Two-Piece Elbow in an Oblong Pipe.   Case I

The only difference to be observed in developing patterns for elbows in oblong pipes, as compared with the same operations in connection with round pipes, lies with the profile or plan.

The plan is to be placed in the same position as shown in the development of patterns for elbows in round pipes, but is placed with the long or narrow side to the view, as the requirements of the case may be.

In Figure 39 is shown the elevation, plan and pattern for a two-piece oblong elbow having semi-circular ends. The narrow side of the oblong is presented to view, with the seam in the center of the long side, as shown by *x* in the plan. Draw the plan and elevation, and divide the upper semi-circular end into a number of equal parts. From these parts draw vertical lines to the miter line *CF*. At right angles to the vertical arm *EF*, draw the stretchout line *x–x*, and thru the points in it draw the usual measuring lines. From the points on the miter line *CF*, draw horizontal lines, intersecting similarly numbered measuring lines, and a line traced thru these points will give the required pattern.

Problem 4.   Case II

In Figure 40 is shown the plan and elevation of a right-angled, two-piece elbow in an oblong pipe, having the long side of the oblong presented to view, as shown by the position of the plan.

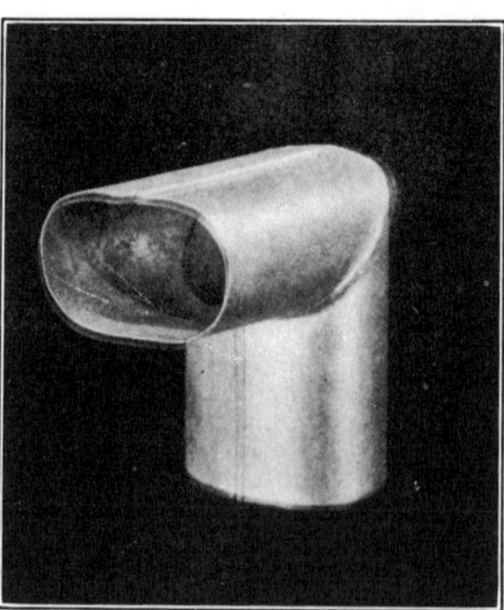

Problem 4.   Two-Piece Elbow, Oblong Pipe.

Draw the plan and elevation and develop the pattern for the vertical arm *EF*. The order of procedure is the same as that given for the drawing in Case I, although the results in the two cases are different in consequence of the position of the profiles. The drawing should present no difficulties to the attentive student.

Problem 5.   Pipe and Roof Flange

Figure 41 shows the method of developing the patterns for a

A
F
Elevation
B
C
D
E
Fig. 40.
Two-piece Elbow
2 3 4 5 6 7 8
x
Plan

2
H
1 3
4
G
4"
4"
m
6"
B
A
C
Fig. 42. Pipe and Roof Flange

3 4 5
2 6
1 7
F
4"
a
b
8"
C
7
6
5
4
3
2
A
e
3 2
4 1
M
J
2
3
4
B
h
G
K
f
10"
9
12"
Fig. 41. Pipe and Roof Flange

A
F
Elevation
B
C
D
E
x 1 2 3 4 5 6 7 7 6 5 4 3 2 1 x
3 4 5
2 6
1 7
x
1 7
Plan
4
Fig. 39. Two-piece Elbow—Oblong Pipe

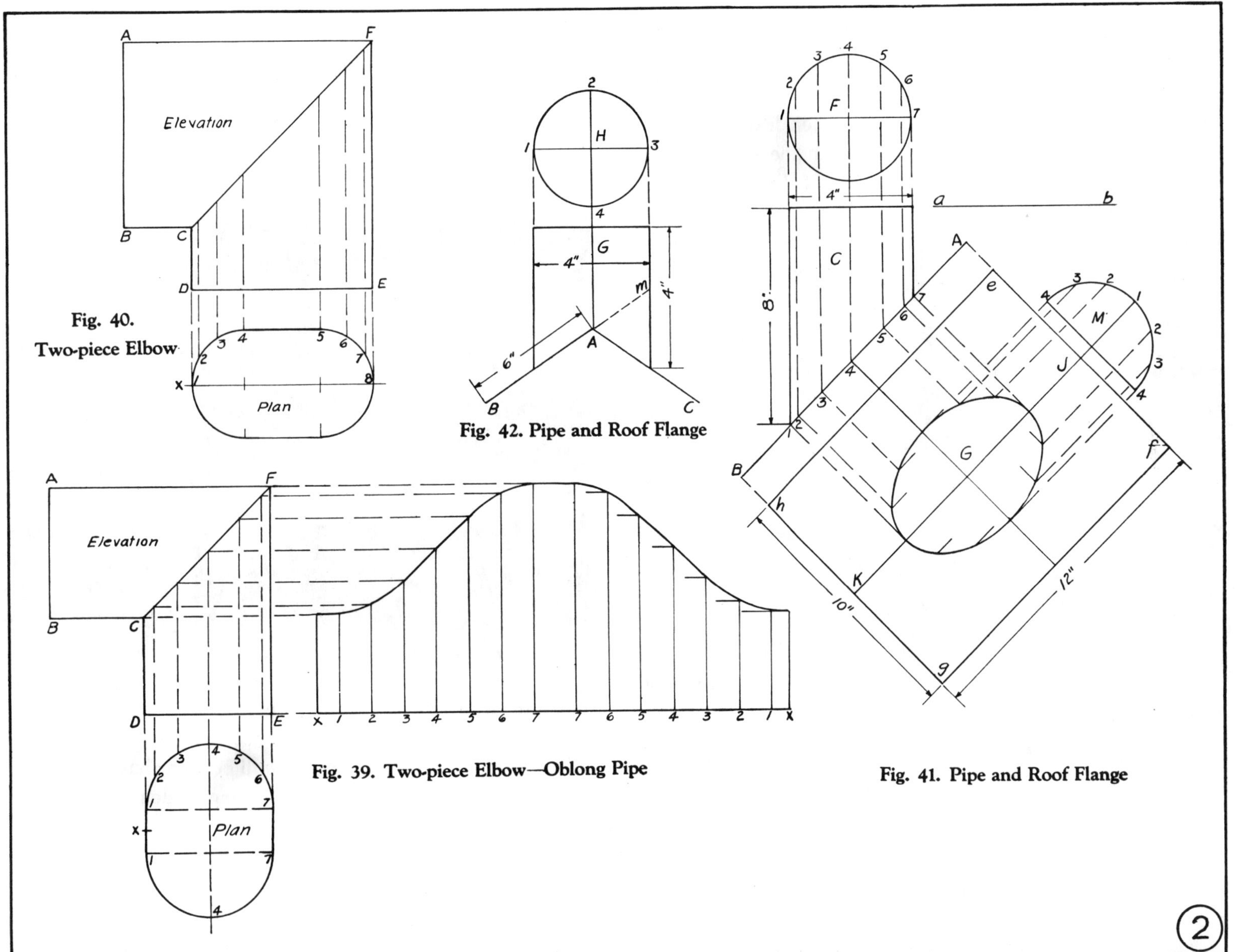

2

pipe and roof flange used by plumbers and sheet-metal workers when flashing around vent pipes and stacks that come thru the slanting sides of a roof.

Draw the roof line *AB* at an angle of 45°. Then draw a side view of the pipe *C*, and immediately above it and in line with the pipe, draw the profile *F*. Divide the upper half of the profile into a number of equal parts, and from these points draw vertical lines to the line *AB*, representing the pitch of the roof. To develop the pattern for the pipe *C*, draw the stretchout line *ab* at right angles to the vertical side of the pipe, and obtain the pattern in a manner similar to the development of the lower arm of the two-piece elbow shown in Figure 36.

The pattern for the opening in the roof flange is shown at *G*, and is developed in the following manner: First, draw lines at right angles to the roof line *AB* from the points *1* to *7*. Then, at right angles to these lines draw the line *J K* thru the center of the roof plate *efgh*. On the line *J K* place one-half of section *F*, as shown by *M*, and divide the half circle into the same number of equal spaces to correspond to section *F*. From these points in *M* draw lines parallel to *J K*, intersecting similarly numbered lines that have been drawn from the points on the line *AB*. A

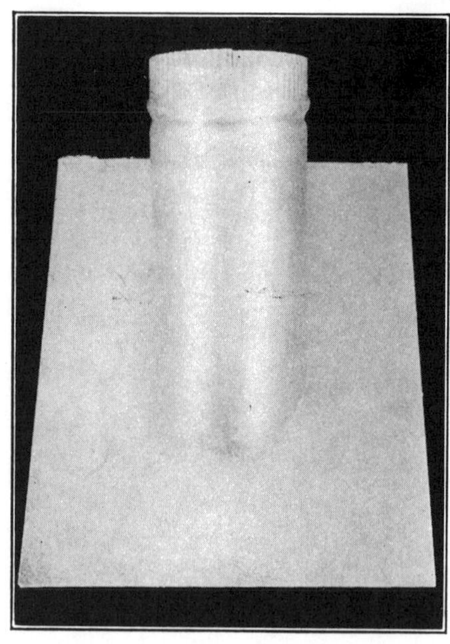

Problem 5.   Pipe and Roof Flange.

line traced thru the points thus obtained will be the pattern for the opening in the roof plate.

Problem 6. Pipe and Roof Flange for Ridge of Roof

In Figure 42, let *ABC* be the length of the roof plate and a section of the roof against which the flange is to fit. Draw an elevation of the pipe *G*, and the profile *H* in their proper position. Then develop patterns for the pipe and the opening in the roof flange in the same manner as described in Problem 5. Since both halves of the opening in the roof flange from the point *A* are the same, both halves of the pattern may be obtained at one operation, the line *BA* may be extended across the pipe to *m* and used in place of *AC*. When the roof line is extended in this manner, it will be seen that the method of developing the pattern for the opening is identical with that shown at *G* in Figure 41.

Problem 7.   Pipe Fitting Over Ridge of Roof.

Problem 7. An Octagonal Pipe Fitting Over the Ridge of a Roof

Applying the method given in Figure 42, develop the pattern for the octagon shaft shown in Figure 43. Let *F* be the section

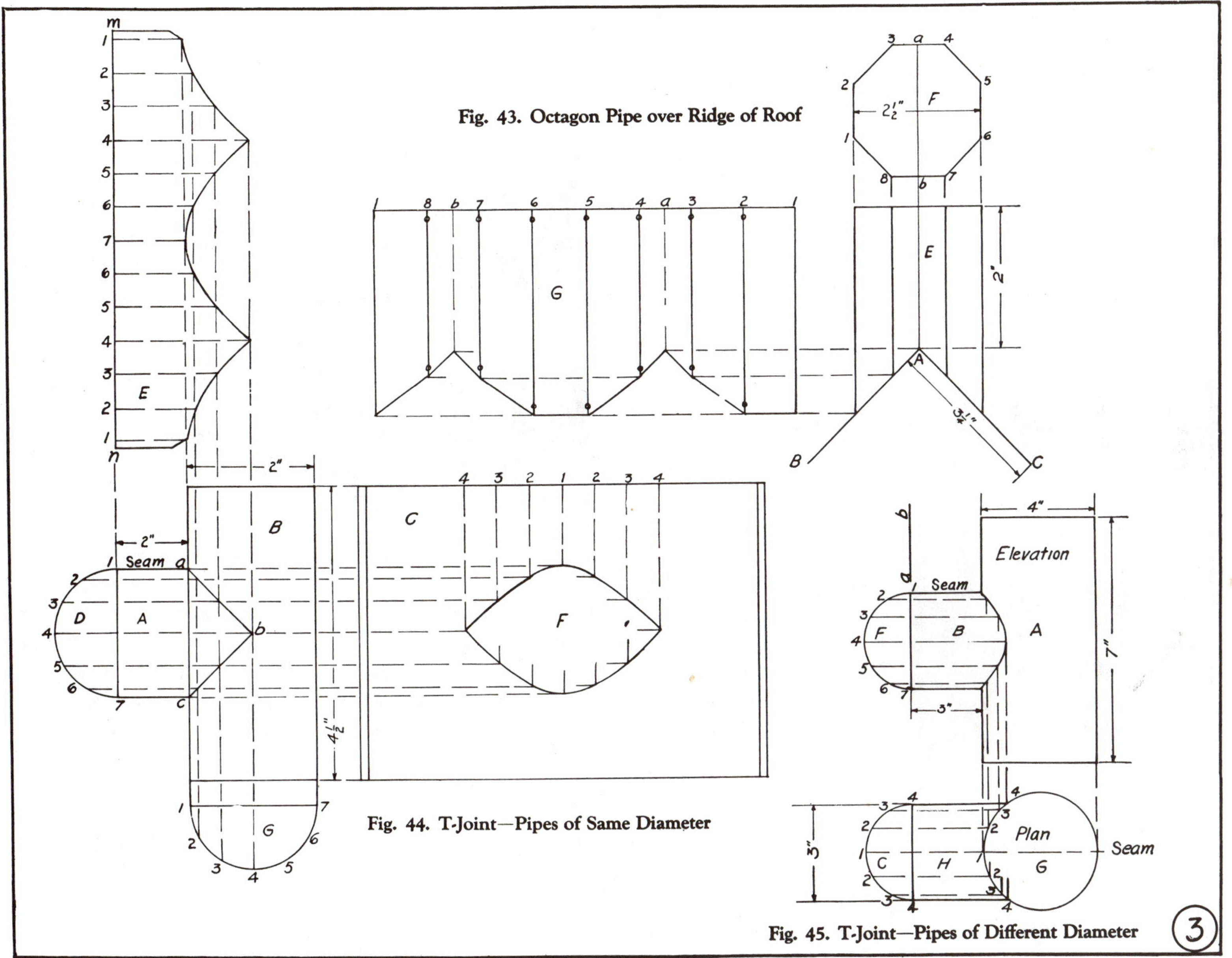

Fig. 44. T-Joint—Pipes of Same Diameter

Fig. 45. T-Joint—Pipes of Different Diameter

and E the elevation of an octagon pipe mitering against a double-pitched roof represented by the lines BA and AC. Draw a vertical line from the point A to the section F, locating the points a and b which are placed in their proper position on the stretchout line.

From these points draw measuring lines which are intersected by a line drawn from the point A, representing the ridge of the roof, as shown in the pattern G.

Problem 8.   T-Joint Between Pipes of Same Diameter

Figure 44 shows the method of developing the patterns for two cylinders of the same diameter intersecting at right angles.

Draw the elevation and place the horizontal pipe A in the center of the vertical pipe B. Then draw the half section of the horizontal pipe A and divide it into a number of equal parts, as shown from 1 to 7 in D. Draw the half section G of the vertical pipe and divide it into the same number of parts as section D, placing the numbers in their proper position as shown. In the half section D the points 1 and 7 are on the top and bottom, while the point 4 is on the long side of the pipe. As both pipes are the same diameter, both halves of pipe A will miter with one-half of pipe B, and when looking down upon the end of the vertical pipe, point 4 will intersect the vertical pipe on the side, as shown by point 4 in half section G.

Problem 8.   T-Joint, Pipes of Same Diameter.

Horizontal lines are now drawn from the points in section D, which are intersected by vertical lines drawn from similarly numbered points in section G. Lines drawn thru these points of intersection will give the miter line. The two pipes being of the same diameter, the miter line is represented by straight lines at an angle of 45°, shown by abc. To obtain the pattern E for the horizontal pipe A, draw the stretchout line mn, upon which step off twice the number of spaces contained in the half section D. From these points draw the usual measuring lines which intersect by vertical lines drawn from similarly numbered points on the miter line abc.

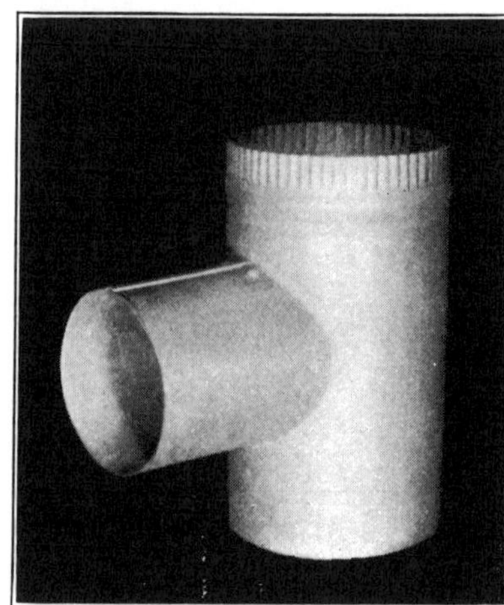

Problem 9.   T-Joint, Pipes of Different Diameters.

A line traced thru these intersections will give the required pattern. The pattern F for the vertical pipe B is simply a rectangle in form, the length being equal to the circumference of the pipe, and the width being equal to the height.

The pattern for the opening F is obtained in the following manner: Locate point 1 upon the upper edge of pattern C, which will be the center of the opening. On each side of point 1 step off the spaces shown from 1 to 4 in the half section G, which will give the length of the opening, and from these points draw vertical lines which are intersected by horizontal lines drawn from similarly numbered points on the miter line *abc* in the elevation. A line traced thru the points thus obtained will give the pattern for the desired opening.

Problem 9. T-Joint Between Pipes of Different Diameters

Figure 45 shows the plan and elevation of a T-joint, the pipes being of different diameters, the horizontal pipe *B* being placed in the center of the vertical pipe *A* at an angle of 90°.

First, draw the plan and elevation, as shown in the drawing. After the outline of the small pipe *H* has been drawn in the plan view, draw the half section *C* and divide it into a number of equal parts. Then draw horizontal lines from these points, intersecting the large pipe *G*, as shown. Next, draw the half section *F* on the end of the small pipe in the elevation, and divide it into the same number of spaces, as the half circle is a duplicate of the half section *C* in the plan.

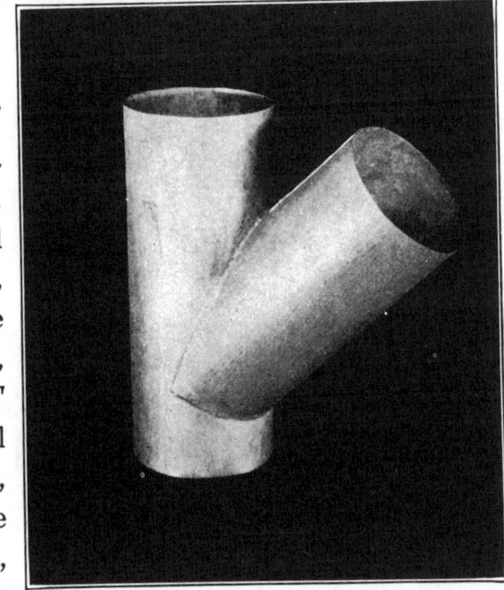

Problem 10. T-Joint, Pipes of Same Diameter at an Angle.

From the points in section *F* draw horizontal lines which are intersected by vertical lines drawn from similarly numbered points on the large circle *G* in the plan. A curved line traced thru these points of intersection will give the miter line between the two pipes. Develop the pattern for the small pipe and the opening in the large pipe in the same manner as explained in the previous problem, shown in Figure 44.

The stretchout of the opening in the large pipe is shown by the figures *1–4–1* in the plan. The spaces being unequal in length, they must be transferred separately to the stretchout line of the pattern.

Problem 10. T-Joint Between Pipes of Same Diameter at an Angle

Figure 46 shows the intersection of two cylinders of equal diameter at an angle of 45°. Let *A* represent the plan of the vertical pipe, and *B* its elevation. Draw the outline of the oblique pipe *C* at its required angle, and place the section *D* in its position, as shown. Space one-half of plan *A*, and section *D* into the same number of equal parts. Draw lines from these points intersecting in the elevation. A line drawn thru the intersections obtained in this manner will give the miter line between the two pipes, shown by *abc* in the elevation. Pattern *E* for the inclined pipe and pattern *G* for the vertical pipe are shown fully developed.

The principles in this problem do not differ from those given in Figure 44. The problems are the same except in the position of the oblique pipe *C*, and the same principles are applicable, no matter what diameter the pipes may have, or at what angle they are joined.

Problem 11. Y-Joint

Figure 47 shows the elevation, partial development and dimensions of a Y-joint, the diameter of each branch being the same.

Draw the elevation, making the arms *AB* at an angle of 90°.

The miter line *ba* is obtained by bisecting the angle *dac* by the line *bg*. The pipes being of the same diameter, a half section of arm *A* shown at *D* is all that is required to obtain the points on the miter lines. Divide the half section *D* into a number of equal parts, being careful to place a point on the quarter circle, as shown by the point *4*. Draw lines from these points to the miter lines *bf* and *ba*, as shown. Then draw the stretchout lines *mn* and develop patterns for the arms *A* and *C*, placing the seams on the short side of both pieces.

Problem 12. Two Square Pipes of the Same Size Intersecting at an Angle

In Figure 48 is shown the intersection between two square pipes of equal size, the inclined pipe *G* being placed in elevation at an angle of 30° to the base line.

Let *ABCD* represent a plan of the vertical pipe placed diagonally, as shown, above which draw the elevation *E*. In its proper position, draw the outline and profile of the inclined pipe in both the plan and elevation, numbering the corners *1-2-3-4*, as shown by *F* and *K*.

From the corner *A* in the plan, draw a horizontal line thru the profile *F*, locating the points *a* and *b*, as shown. These points must also be placed in their proper position in profile *K*, and on the stretchout line when developing the pattern for the pipe *G*. Develop the patterns for the vertical pipe *E* and the inclined

Problem 12. Square Pipes of Same Diameter Intersecting at an Angle.

pipe *G*; also the pattern for the opening in the vertical pipe. The principles explained in Figure 46 for the intersection of round pipes are also applicable for pipes that are square, rectangular, elliptical and oblong in form.

Problem 13. Two Round Pipes of Unequal Diameters That Intersect Irregularly

Figure 49 shows the elevation, plan and patterns of two cylinders of different diameters intersecting at an angle of 45°. The position of the two pipes is such that the outline *7-7* of the smaller pipe in the plan view is tangent to the circle that represents the large pipe. First, draw the plan *A* of the large pipe and the profile *B* of the smaller pipe in their proper position.

Space the profile *B* into a number of equal parts, and from these points draw lines intersecting the large pipe in the plan at *A*.

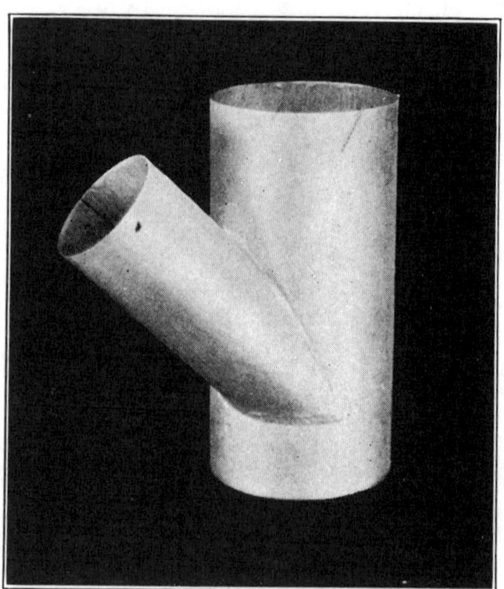

Problem 13. Two Round Pipes of Unequal Diameters Intersecting Irregularly.

Draw the elevation of the two pipes, as shown at *F* and *H*, and describe the circle that represents the profile of the small pipe, as shown at *C*. Divide the profile *C* into the same number of equal spaces as the profile *B*, and from these points draw lines parallel to the inclined arm indefinitely. From the points in

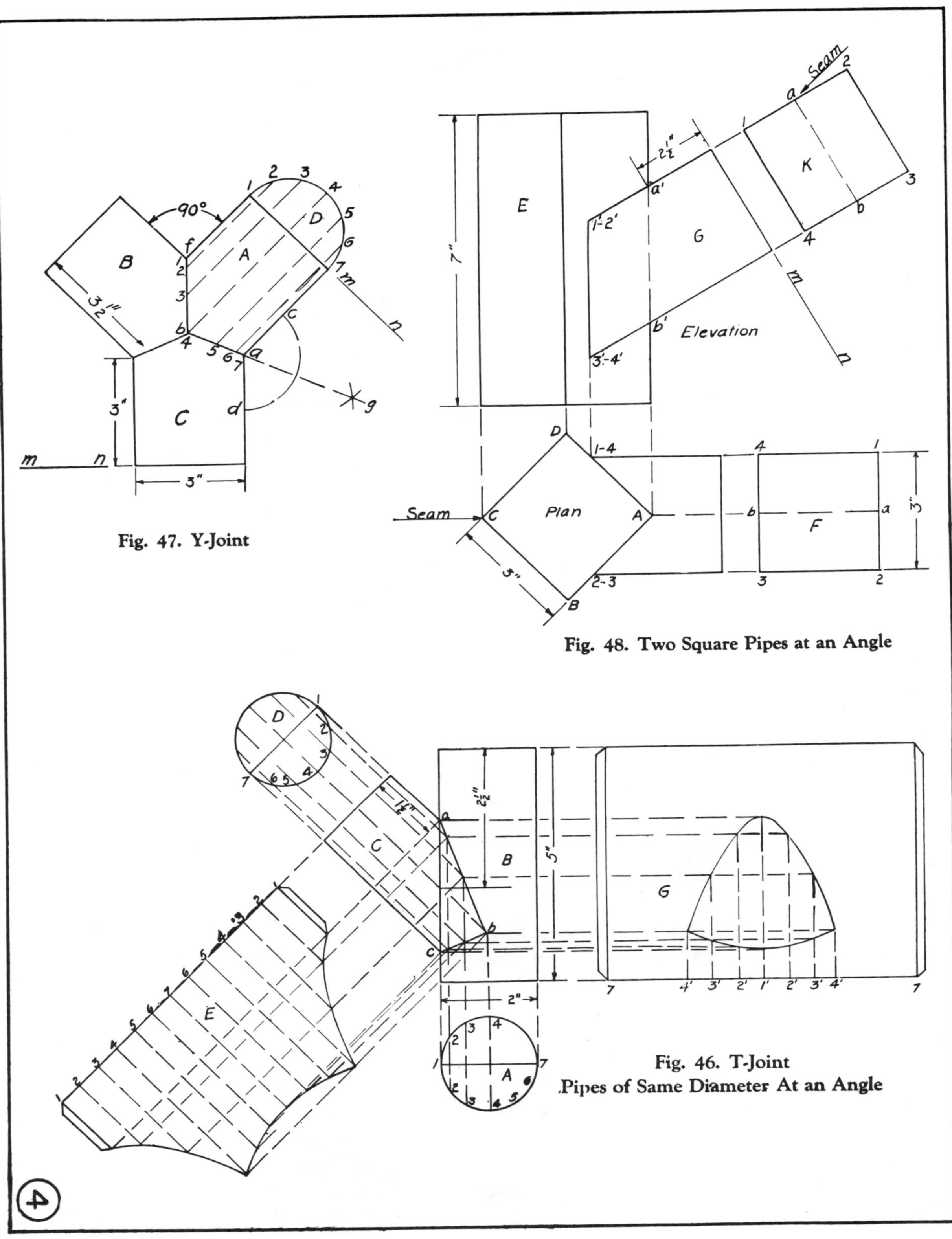

90°
B
D
A
f
C
m
n
g
3"
3"
Fig. 47. Y-Joint
Seam
a
E
G
K
b
Elevation
m
n
D
Plan
A
C
B
Seam
F
a
b
Fig. 48. Two Square Pipes at an Angle
D
C
E
B
A
G
Fig. 46. T-Joint
Pipes of Same Diameter At an Angle

circle *A* of the plan, draw vertical lines, intersecting the oblique lines drawn from points in profile *C*. A line drawn thru these points of intersection will be the miter between the two pipes. Note the position of the points on the circle *B* that represents the profile of the smaller pipe in the plan at *1, 2, 3*, etc., and their corresponding location at *1, 2, 3*, etc., in the profile *C* in the elevation.

For the pattern of the small pipe, proceed as follows: Draw the stretchout line *ab* at right angles to the arm *H*, and draw the measuring lines, as shown, numbering them to correspond to the spaces in profile *C*. The seam has been placed on the short side of the pipe at point *1*. Therefore, commence numbering the stretchout line with *1*. From the points *1, 2, 3*, etc., on the miter line, draw lines at right angles to the arm *H*, intersecting similarly numbered measuring lines.

A line traced thru these points will give the required pattern shown at *G*.

The development of the large cylinder is shown at *K*. Draw the stretchout line and make *x′–x′* equal in length to the circumference of the circle shown at *A* in the plan.

Reference to the plan shows that the small pipe intersects the surface of the larger from points *1* to *7* on the large circle, which

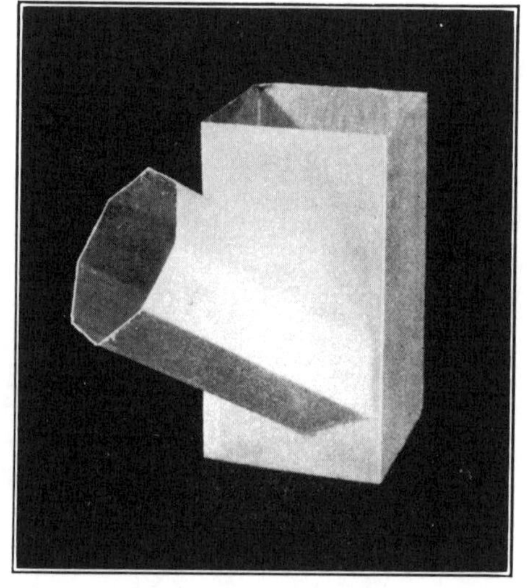

Problem 14.   Octagonal Pipe Intersecting a Square Pipe.

will be the stretchout for the pattern of the opening in the large pipe. For the pattern of the opening, locate the point *7* on the stretchout line *x′–x′*, the distance from *x′* to *7* being one-quarter of the circumference of the large pipe. Then take the divisions *7–6, 6–5, 5–4*, etc., on the large circle in the plan, and place them on the stretchout line *x′–x′*, as shown.

From these points draw vertical lines which are intersected by horizontal lines drawn from similarly numbered points in the elevation.

A line traced thru these points of intersection will give the irregular outline of the pattern, shown at *P* in the drawing.

Problem 14.   An Octagonal Pipe Intersecting a Square Pipe Obliquely

Figure 50 shows the plan, elevation and intersections of a square and octagonal pipe, the octagonal pipe being placed to one side of the center of the square pipe and inclined at an angle of 30° to the base line. Draw the plan of the square pipe, as shown at *ABCD*. Locate the point *3–4* one-half inch from the corner *B*, and draw the profile of the octagon pipe in the position shown at *E*.

Draw the elevation and profile *F*, as shown, placing the numbers on the profile in their proper position. It is necessary to use extreme care in numbering the points on the two profiles, in order that the position of the points on the profiles may be located in the same corresponding position with regard to one another. From the center of the square pipe shown at *A* in the plan, draw a line intersecting the profile *E* at *x* and *y*.

The points *x* and *y* must be placed in their proper position in the profile *F*, also on the miter line and stretchout line, before developing the patterns. This problem introduces no new element in the development of solids by parallel lines. The prin-

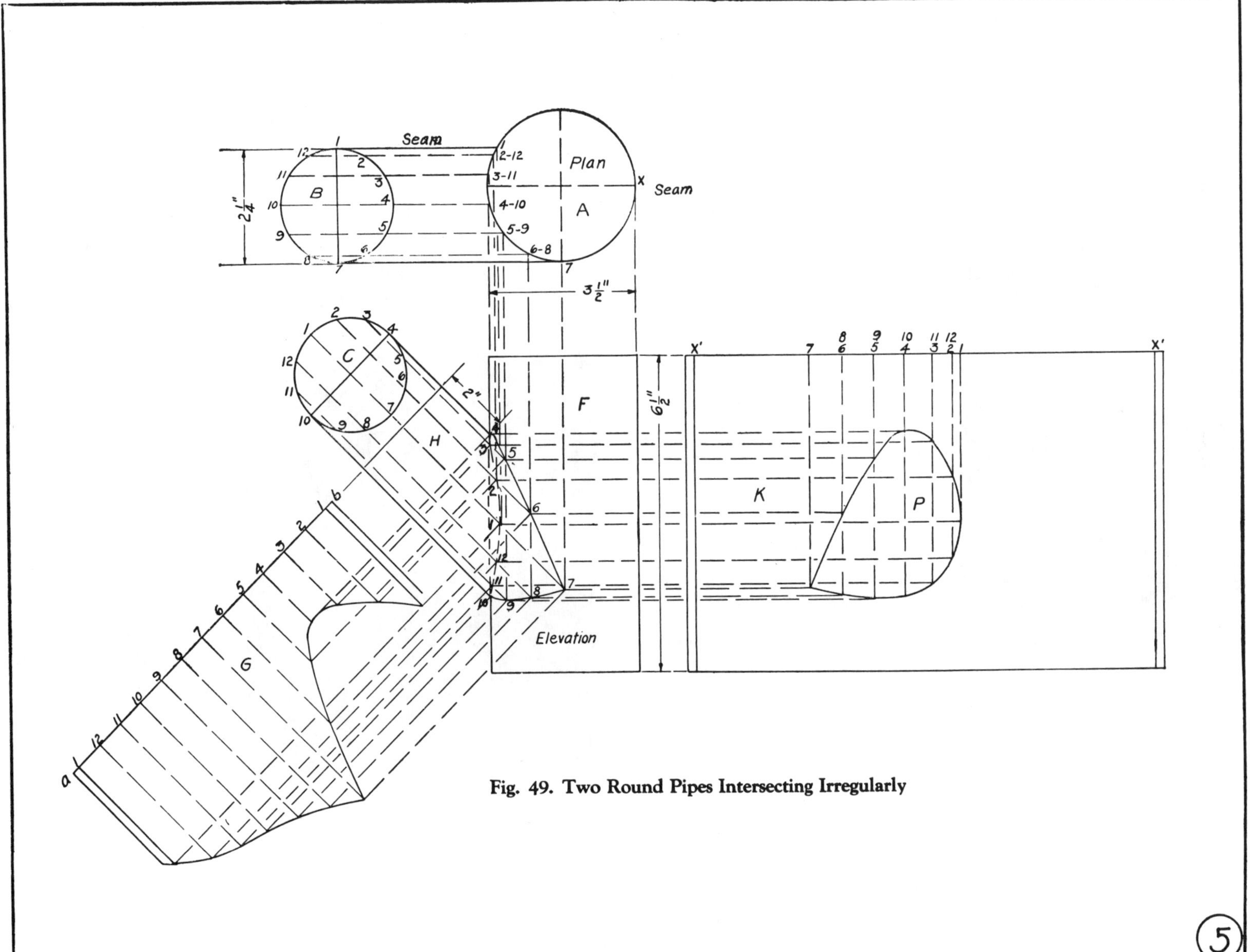

Fig. 49. Two Round Pipes Intersecting Irregularly

ciples are the same as those governing the development of the round pipes in the previous problem. Complete the problem by drawing the miter lines, and develop the patterns for the square and octagon pipes. Also a pattern for the opening in the square pipe.

Problem 15. A Rectangular Pipe Intersecting a Round Pipe Obliquely

Problem 15. Rectangular Pipe Intersecting a Round Pipe.

In Figure 51 is shown the plan and elevation of a rectangular pipe intersecting a cylinder obliquely. Draw the plan and elevation, placing the oblique pipe C in the elevation at an angle of 45° to the base line.

Draw the profiles of the rectangular pipe in the position shown at F and G.

Develop the patterns for the rectangular pipe C, and the opening in the round pipe A, as shown at B, in accordance with principles already explained.

Problem 16. Four-Piece 90° Elbow

Figure 52 shows the method of obtaining the patterns for a four-piece 90° elbow having a diameter of 5 inches; the length of the radius for the inner curve of the elbow being 3 inches.

Draw the right angle ABC, and on the line BC lay off a distance of 3 inches from B to n. With B as center and Bn as radius, describe the quarter circle mn, which gives the required curve for the throat.

Make nc the diameter of the elbow, and with BC as radius and B as center, describe the outer arc CA.

The miter lines of the elbow, shown by EFG, are obtained by dividing the outer quarter circle AC into equal parts one less in number than the pieces required in the elbow; in this case, into three parts, shown by Ae, ee and ec. Bisect each of these spaces, and lines drawn from these points to the center B will give the joint or miter lines of the elbow. This method can be used to obtain the miter line for elbows at any angle having any desired number of pieces. Next, draw the half section D, and divide it into a number of equal spaces, and from these points draw vertical lines intersecting the miter line BG in the elevation.

To develop the pattern for the first section of the elbow, draw the stretchout line HR, upon which place twice the number of spaces contained in half section D. From these points draw the usual measuring lines, which intersect lines drawn from similarly numbered points on the miter line BG. Thru the points thus obtained, trace the irregular curve of the pattern, as shown by gLg. This irregular curve is the only one needed, and is used in laying out the patterns for the other sections of the elbow.

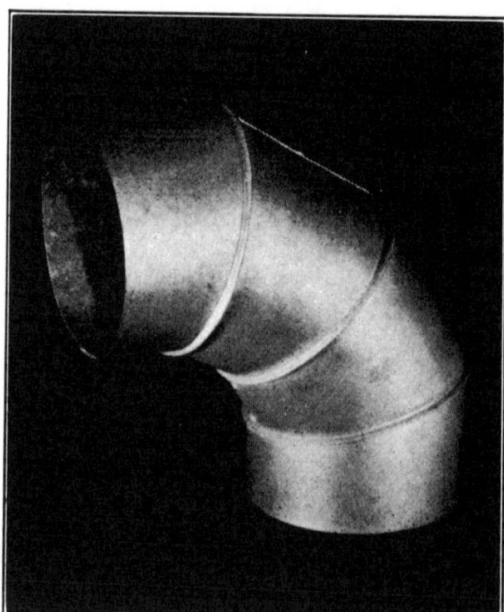

Problem 16. Four-Piece 90° Elbow.

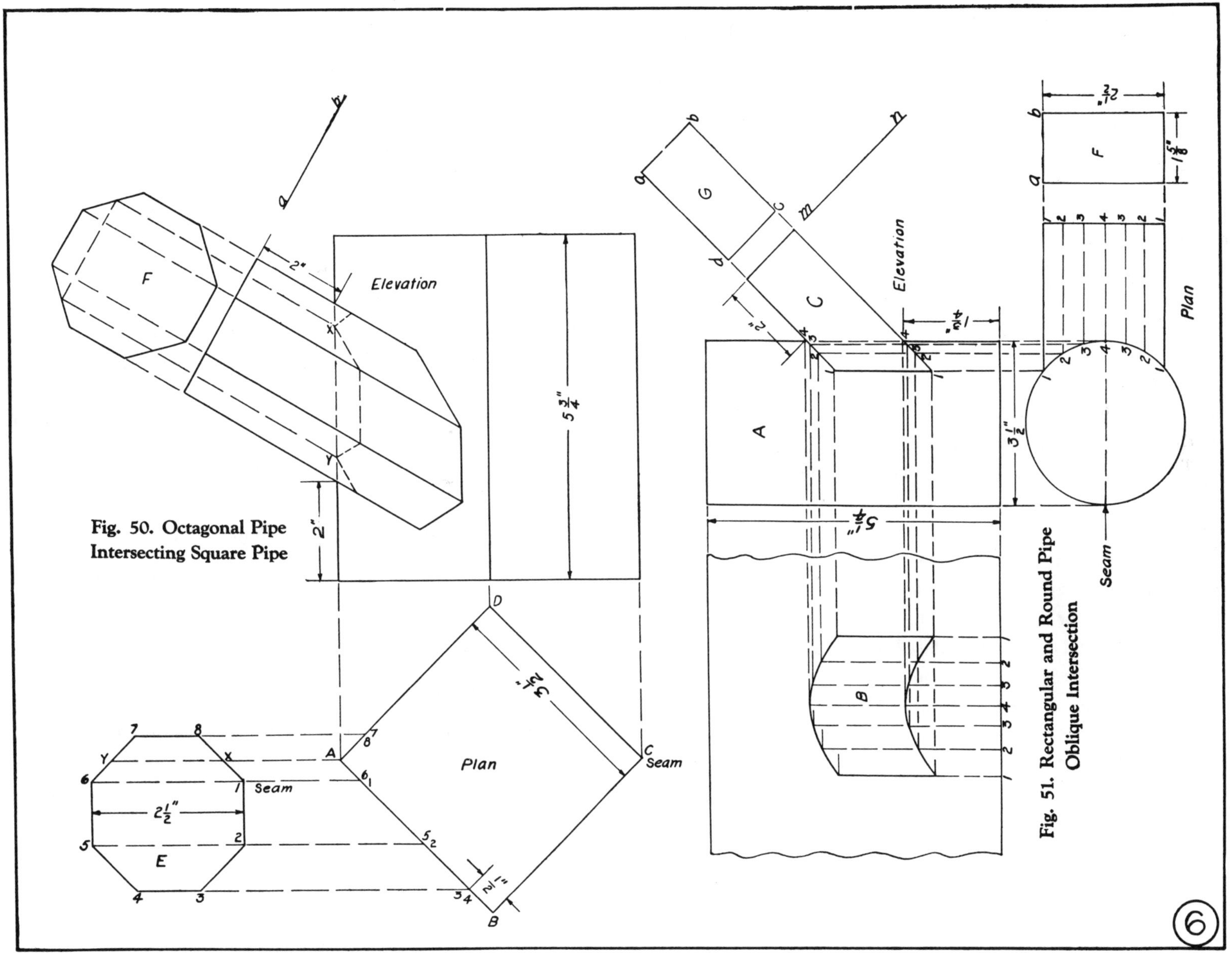

Fig. 50. Octagonal Pipe Intersecting Square Pipe

Fig. 51. Rectangular and Round Pipe Oblique Intersection

⑥

The patterns for the sections *2*, *3* and *4* are usually laid out in the following manner: As the patterns are all of equal length, from point *1* on each end of the stretchout line *HR*, draw the vertical lines *1m*. Next, take the wide side of section *2*, as shown by *GF* in the elevation, and mark this dimension on the lines *1m* at both ends of the pattern, as shown by *gb*.

Then take the throat width of piece *3* and place it as shown by *ba*. The length of long side or top of piece *4* is placed on the same lines, as shown by *ma*. The rivet holes are spaced, as shown by the small circles on the lines *1m*.

This completes the drawing, as the patterns are now laid out directly on the metal in the following manner: Place the drawing upon the metal, and, by means of a prick punch, transfer pattern *L* to the metal by pricking along the irregular curve. The centers of rivet holes and the width of sections are also pricked thru the drawing in the usual manner. Pattern *L* is now cut from the metal, after which the metal pattern is reversed and the curved edge placed on the points *bb*. Using a scratch awl, the irregular curve is scribed upon the metal, which completes the pattern for piece *2*, shown at *T*.

The patterns for pieces numbered *3* and *4* are completed by placing the curved edge of the metal pattern *L* on the points *aa*; then scribe the irregular curve on the metal, and draw a line from *m* to *m*, completing the patterns *P* and *K*. This method of grouping the patterns places the seams opposite each other and allows the patterns to be cut without waste of material.

Problem 17.   Five-Piece 60° Elbow

Figure 53 shows the elevation of a five-piece elbow, which, when completed, should have an angle of 60°, the inner curve or throat being described with an 8-inch radius.

First, draw the required angle *ABC*; next, on the line *AC*, measure off a distance of 8 inches from *A* to *G*. With *AG* as radius and *A* as center, describe the arc *GF*.

Make *GC* equal 5 inches, and with *AC* as radius, describe the arc *CB*, which is divided into four equal spaces, one less in number than the pieces in the elbow. These spaces are shown by *Be*, *ee* and *eC*. Bisect these spaces, as shown at *m, m, m, m,* and lines drawn thru these points from the apex will give the miter line for each section of the elbow. The end pieces *1* and *5* may be made any length, but the length of the heel and throat of the middle sections should be taken from the elevation, as shown by *aa* and *bb* in section *3*, and cannot be changed when once the arc *GF* has been described on the drawing. Draw the half section *K*, complete the elevation, and develop the pattern for piece *1*

Problem 18.   Three-Piece 90° Elbow

Figure 54 shows the elevation of a three-piece 90° elbow for which the pattern of piece *1* is to be obtained in accordance with principles already explained in Problem 16.

Problem 19.   Profiles of Elbows

Figure 55 shows the profiles or sections of a square, rectangular, and oblong elbow, shown at *A*, *B* and *C*. Develop the pattern for a 2-, 3- and 4-piece elbow having a 3-inch throat radius and an angle of 90°. The patterns for the end pieces only are required.

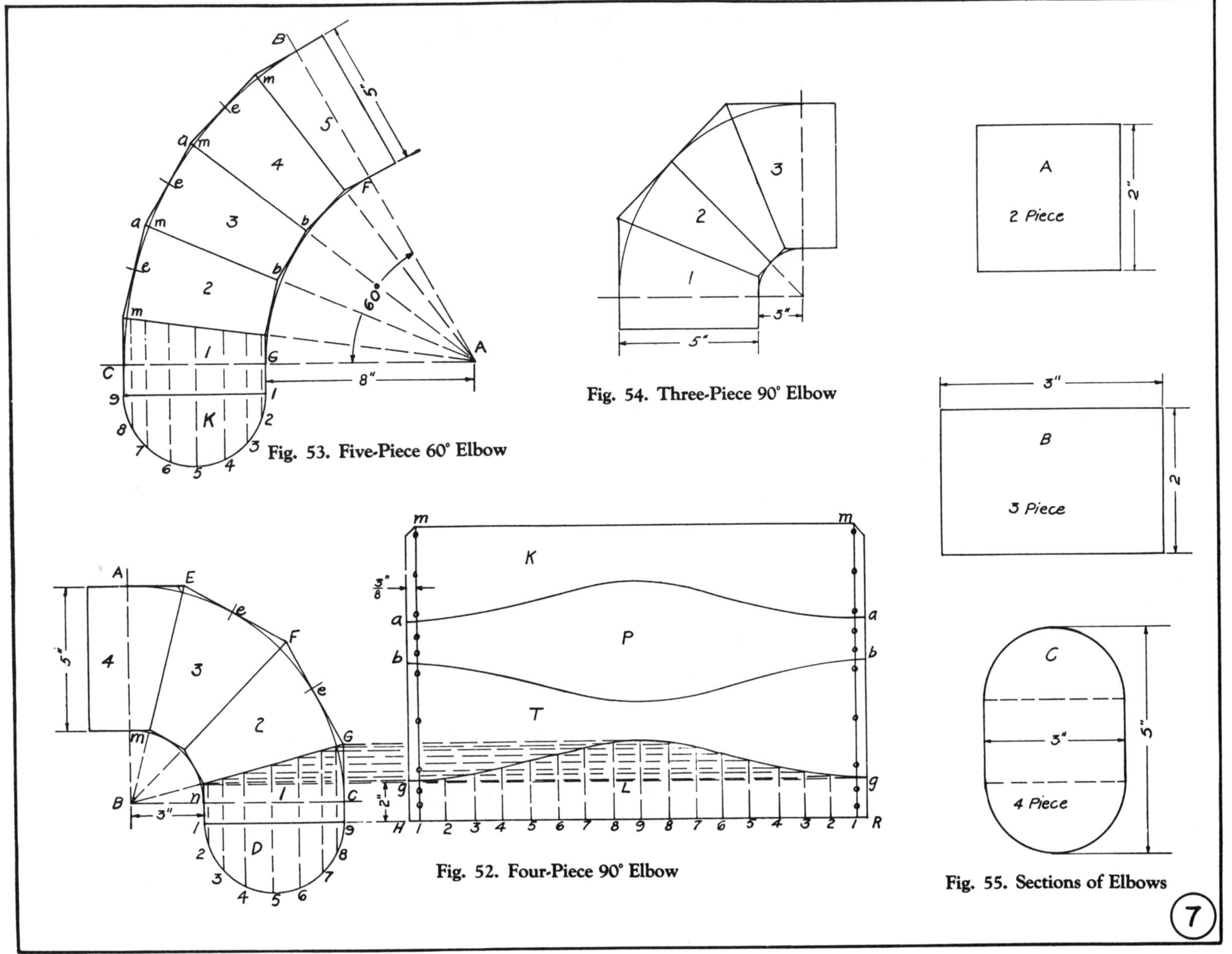

Fig. 53. Five-Piece 60° Elbow

Fig. 54. Three-Piece 90° Elbow

Fig. 52. Four-Piece 90° Elbow

Fig. 55. Sections of Elbows

CHAPTER V

Practical Cornice and Gutter Problems

The patterns for miters between sheet-metal moldings are developed by the parallel-line method described in the previous chapter, and in order to illustrate the application of the principles as applied to moldings, a number of practical problems are presented.

Problem 20. Roman Moldings

In classical architecture, Greek and Roman moldings are employed. The outlines of Greek moldings which, in nearly every instance, are found to follow the curves of the conic sections, generally the parabola or the hyperbola.

Roman moldings are nearly always formed from the arcs of one or more circles, and are chiefly used in sheet-metal cornice work. Figure 56 shows a series of Roman moldings, and demonstrates a method by which each may be geometrically drawn.

The *torus* molding, shown at A, is half round and here shown between two fillets. It is a semi-circle described from the center C, the bisection of the line ab.

The *cavetto*, or *cove*, shown at B, is a concave molding whose profile is a quarter circle. The center c is found by extending the lines ab and dc until they intersect.

The *ovolo*, or *quarter round*, shown at C, is a convex molding with a quarter-circle profile; the center b is obtained in a similar manner as the previous molding.

The *cyma recta*, known as the *ogee*, shown at D, is made up of two quarter reverse circles tangent at m. The centers g and e are found by bisecting the lines ab and cd, as shown.

The *cyma reversa*, shown at E, is the reverse of the cyma recta, which is concave above and convex below; while the cyma reversa is convex above and concave below. The method of construction is similar except the centers g and h are at the top and bottom, as shown.

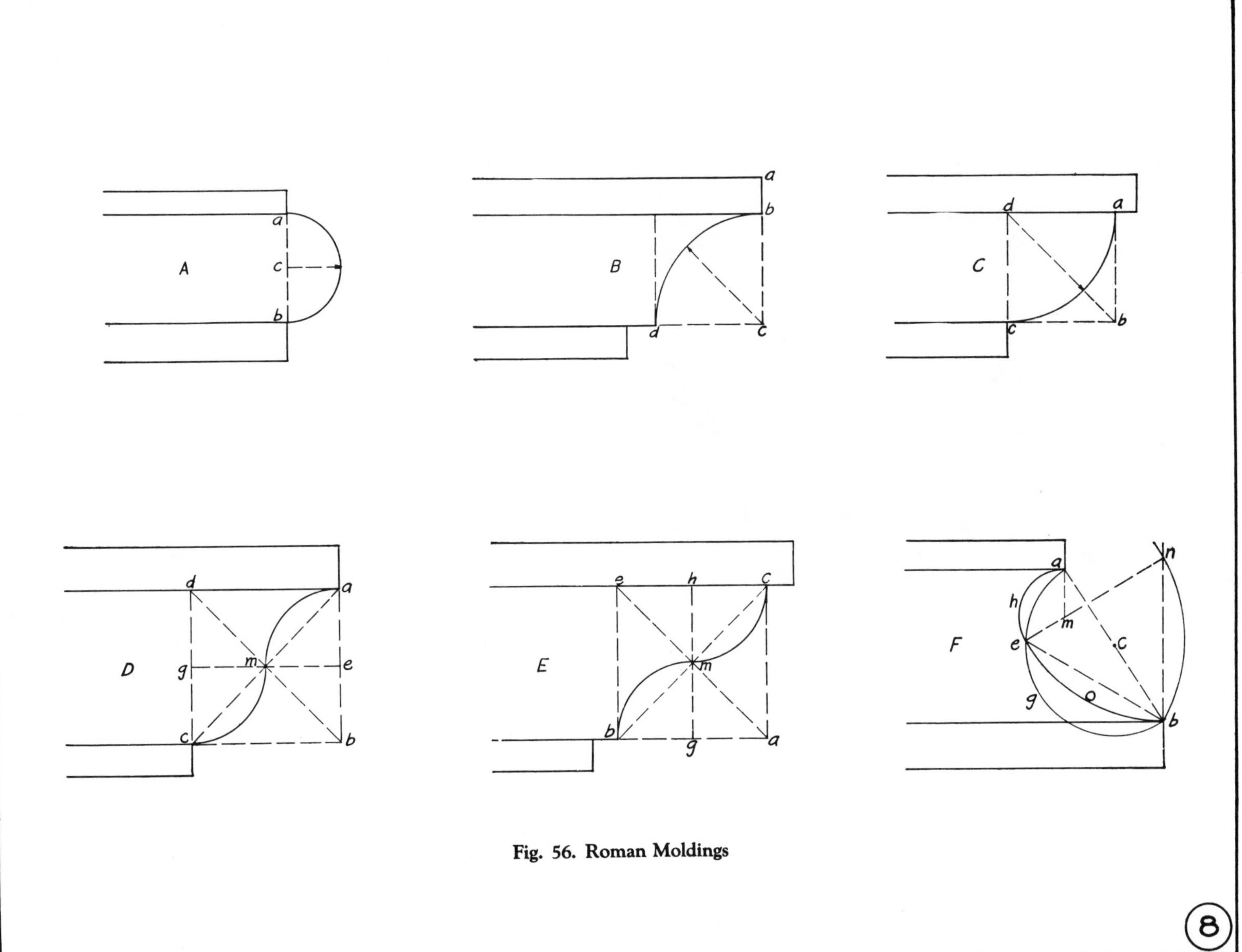

Fig. 56. Roman Moldings

The *scotia*, shown at *F*, is drawn as follows: Having given the points *a* and *b*, draw the line *ab* and bisect *ab* at *c*. From *c*, with a radius equal to *ca*, describe the semi-circle *agb*. From *b*, draw the line *be* at an angle of 30° to the base line, cutting the arc at *e*.

From *b*, erect the perpendicular *bn*, and with *e* as center and *eb* as radius, describe an arc, cutting the perpendicular at *n*.

Draw *ne* and drop a perpendicular from *a*, intersecting *ne* at *m*. With *m* as center and *ma* as radius, draw the arc *ahe*. With *n* as center and radius *ne*, describe the arc *eob*, completing the molding.

Problem 21.   Square Return Miter.

Problem 21.   Square Return Miter

Figure 57 shows the short method of obtaining the pattern for a square return miter.

This method can only be used when the miter is one of 90°; that is, a square miter. Let *C* represent the profile of a molding which is drawn to a scale of 4 inches to 1 foot. Divide the quarter round into a number of equal spaces. Number these points—also the corners of the molding—as shown by the figures *1* to *15*. The stretchout of the molding may be conveniently placed either above or below the drawing of the profile. In this

case, it was drawn below, as shown by the vertical line *AB*, from which horizontal measuring lines are drawn and intersected by vertical lines drawn from similarly numbered points on the profile *C*. A line traced thru the points thus located will complete the pattern, shown at *G*. The edge lines along which bends are to be made are designated by circular indicators in the usual way.

Outside and Inside Miters

Figure 58 shows the sketch of a roof plan to illustrate the difference between an outside miter and an inside miter. Miters for the outer and inner angles of a roof are called outside and inside miters, and are placed as shown in the sketch.

Pattern *G*, shown by *ABce* in Figure 57, is the pattern for an outside miter, while the opposite cut, shown by *cghe* is the pattern for an inside miter. Both patterns are produced by a single miter cut, and this is also true when developing patterns for miters at any angle.

Problem 22.   Butt Miter

Figure 59 shows the elevation of a horizontal molding that butts against a mansard or other pitched roof, and illustrates the principle applicable to butt miters, whether the molding butts against a plain or curved surface in the elevation. Let *E* represent the side elevation of a cornice, which is drawn to the scale of 4 inches to 1 foot. Draw the profile *AB* and divide it into equal spaces, as shown. From these points on the profile, draw horizontal lines parallel to the lines of the molding until they intersect the roof line *CG*.

At right angles to the lines of the molding, draw the stretchout line *ab*, either above or below the profile. Draw the usual measuring lines which are intersected by vertical lines drawn from the various intersections on the roof line *CG*. A line drawn

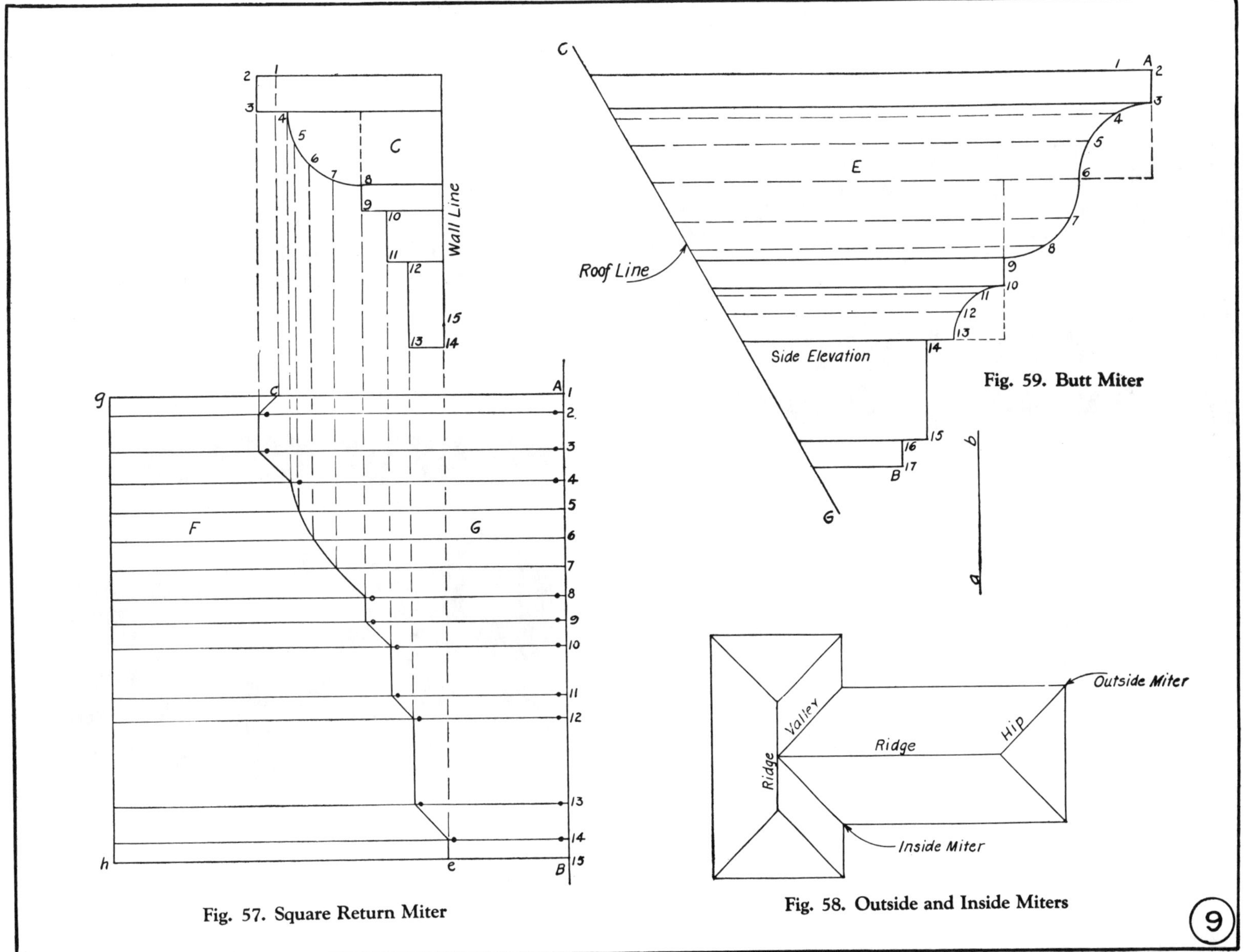

Fig. 57. Square Return Miter

Fig. 58. Outside and Inside Miters

thru these points will be the pattern for the butt miter. Complete the pattern by developing a square return miter cut on the end *AB*.

Problem 23.   An Oblique Return Miter

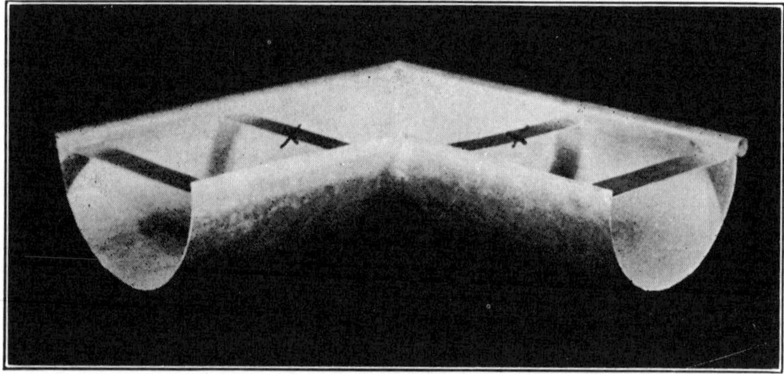

Problem 25.   Eave-Trough Gutter.

Figure 60 shows the method of obtaining the pattern for an octagon return miter, and is applicable for miters at any angle. Patterns for return miters at other than a right angle must be developed from a plan of the molding. In this figure, *A* is the profile of a molding which is drawn to a scale 4 inches to 1 foot. To draw a plan view of the molding, extend the wall line of profile *A*, and by the aid of a 45° triangle, draw the octagonal angle of 135°, as shown by *emn*, which will represent the wall line in plan *B*. Bisect the angle *emn* and draw the miter line *Gm*. Divide the curve in the profile into any convenient number of spaces and set off the entire stretchout of the profile on the line *ab*, drawn at right angles to the lines of the plan. From all points on the profile *A*, draw vertical lines intersecting the miter line *Gm* in the plan. Measuring lines are drawn from the points on the stretchout line *ab*, which are intersected by horizontal lines drawn from similarly numbered points on the miter

line *Gm*. A line traced thru these intersections will complete the pattern, as shown at *C*. The opposite cut of the outside miter shown at *E*, is used should an inside miter be required.

Problem 24. Butt Miter Against a Surface Oblique in Plan

Figure 61 shows a horizontal molding at an angle that butts against a flat surface, as in the case of a bay window cornice that miters against a vertical wall. In this figure, which is drawn to a scale 4 inches to 1 foot, *cemy* represents the face of the wall against which the molding *A* is placed, and *GB* represents the vertical wall against which the butt miter is made.

Bisect the angles to find the miter lines. At right angles to lines of the molding draw the stretchout line *ab*, and develop patterns for moldings *H* and *K*. Compare the view here shown

Problem 26.   Molded Face Gutter.

with that given in Figure 60, and it will be seen that the process of development differs but little from that already explained.

Problem 25.   Eave-Trough Miter

Figure 62 shows the section and pattern for a half-round eave-

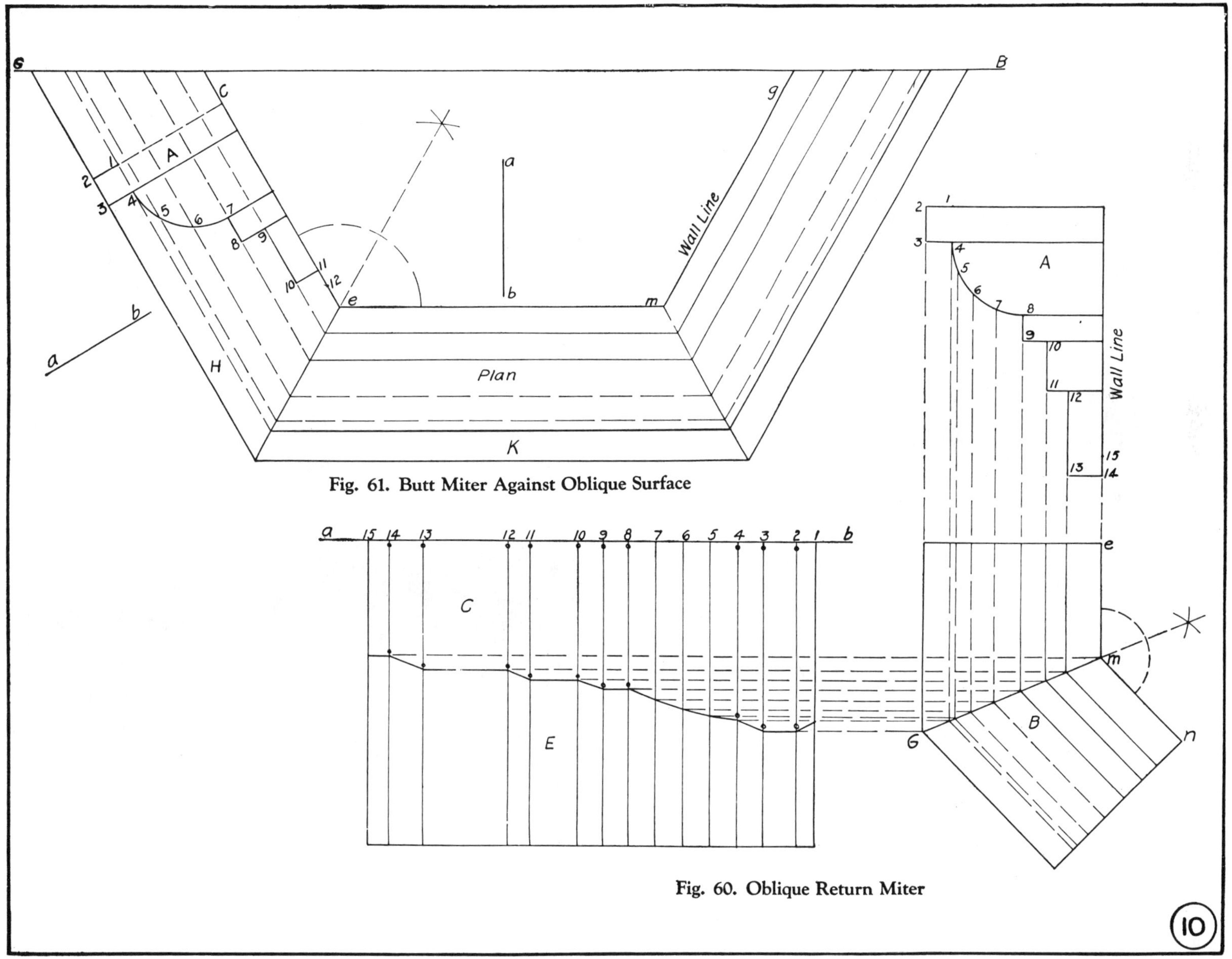
C
A
B
Wall Line
a
b
Plan
e
m
H
K
Wall Line
A
e
a
b
Fig. 61. Butt Miter Against Oblique Surface
a   15 14 13   12 11   10 9 8 7 6 5 4 3 2 1   b
C
E
G
B
m
n
Fig. 60. Oblique Return Miter
10

trough miter having an angle of 90°, which is drawn to a scale of 6 inches to 1 foot.

The gutter is in the form of a half circle having a bead along one of its edges, as shown by the smaller circle.

Draw the end view of the gutter in the position shown at *A*. Space the outline of the end view, as in former problems, and locate the points shown from *O* to *14*.

Draw the stretch-out line *ab*, and develop the pattern for a square return miter in the same manner as described in Problem 21. Pattern *B* is the pattern for an outside miter, while the opposite cut shown at *C* is the pattern for an inside miter.

Problem 27. Ogee Gutter.

Problem 26. Molded Face Gutter

In Figure 63 is shown a drawing which scales 4 inches to the foot, giving the section of a molded face gutter, for which a square return miter pattern is to be developed.

Problem 27. Ogee Gutter

Figure 64 shows the end view or section of an ogee gutter, which is drawn to a 4-inch scale. Develop the pattern for a return miter having an angle of 135° in the plan view. The method of development is the same as that shown in Figure 60.

Problem 28. Molded Roof Gutter

Figure 65 shows the section of a roof gutter which is drawn to a scale of 4 inches to 1 foot.

Let *1, 2, 3, 4, 5* represent the profile of the face, and the line *5, 6* the back of the gutter. A band iron brace is shown at *ce*, which is bolted to the gutter at *a*, and secured to the roof by nailing at *b*. Develop the pattern for a return miter having an angle of 90°.

Problem 29. Octagonal Gutter

In Figure 66 is shown a drawing 4 inches to 1 foot, giving the profile of an octagonal gutter.

Draw the section and plan view and develop the pattern for a return miter having an angle of 120°.

Problem 30. Quarter-Circle Gutter

The section of a quarter-circle gutter is shown in Figure 67.

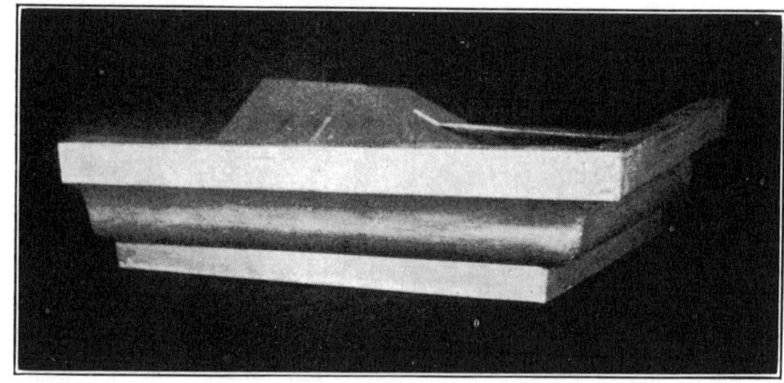

Problem 28. Molded Roof Gutter.

Obtain the miter line in the plan view and develop the pattern for an outside return miter at an angle of 150°. The profile in the drawing is shown one-third full size.

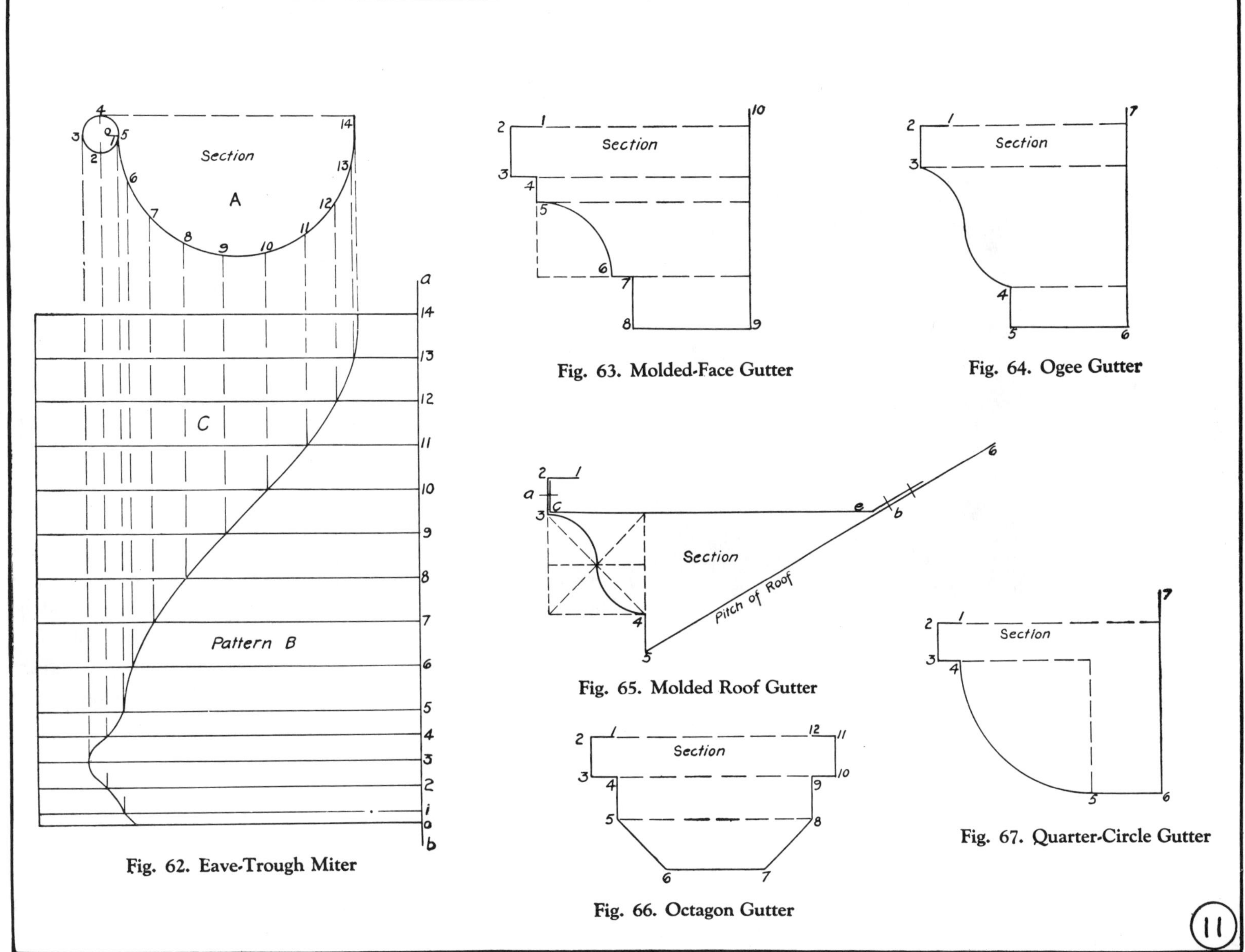
Section
A
Pattern B
C
Fig. 62. Eave-Trough Miter
Section
Fig. 63. Molded-Face Gutter
Section
Fig. 64. Ogee Gutter
Section
Pitch of Roof
Fig. 65. Molded Roof Gutter
Section
Fig. 66. Octagon Gutter
Section
Fig. 67. Quarter-Circle Gutter

Problem 31.   Square-Face Miter (Short Method)

The method of developing the patterns for a square-face miter is shown in Figure 68, and the drawings are made to a scale of 4 inches to the foot.   This process is employed when developing the patterns for miters in gable moldings, picture frames, window and door frames, and panel moldings. The method of development is similar to that described for the square return miter (Figure 57). The only difference is in the position of the stretchout line *ab*.   In this case, the stretchout line is placed in a horizontal position at the left of profile *A*, with the pattern shown at *C*, while the stretchout for the square return miter (Figure 57) is placed in a vertical position below the profile.

Problem 31.   Square-Face Miter.

When developing patterns for square miters, the stretchout line must be placed in its proper position, or, instead of having a face miter, as shown in Figure 68, the draftsman will have a return miter, as shown in Figure 57.

Problem 32.   Octagon-Face Miter

Figure 69 shows the development of an octagon-face miter. The patterns for face miters at other than a right angle are de-veloped by the long method, and the miter line is found in the elevation.   In this problem the elevation is shown at the right of profile *A*.   First, draw the profile *A*, then draw the required angle *BGE*, which is bisected to obtain the miter line *GF*.   From the various points on the profile *A*, draw horizontal lines inter-secting the miter line *GF*, as shown.   Draw the stretchout line *ab* at right angles to the horizontal lines of the molding.

Upon this line place the stretchout of the profile *A*, and de-velop the pattern in the usual manner, as shown by *amnb*.

The method given in this problem is applicable for miters at any angle, and the miter line will be found in the elevation.   In the case of a return miter, the miter line will be found in a plan view, as shown in Figure 60.

Problem 33.   Miter Between a Gable and Horizontal Molding

Figure 70 shows another form of face miter in the style of a gable molding, shown at *AB*, which miters with a horizontal molding *C* that butts against an inclined surface, shown by the line *GF*.   Draw the elevation, which is shown one-third size in the drawing.   The miter lines *RD* and *HK* are found by bisecting the angles in the usual manner.   Place the profile *E* in position and develop patterns for the inclined molding *B* and the horizontal molding *C*.   The principles used in developing this problem are similar to those given in Problem 32.

Problem 34.   Panel Miter

Figure 71 shows the elevation of a rectangular panel which is drawn to a 3-inch scale.

Draw a section of the panel mold *A*, as indicated by the shaded portion of the elevation.   Divide the cove into a number of equal spaces and number each point on the profile.

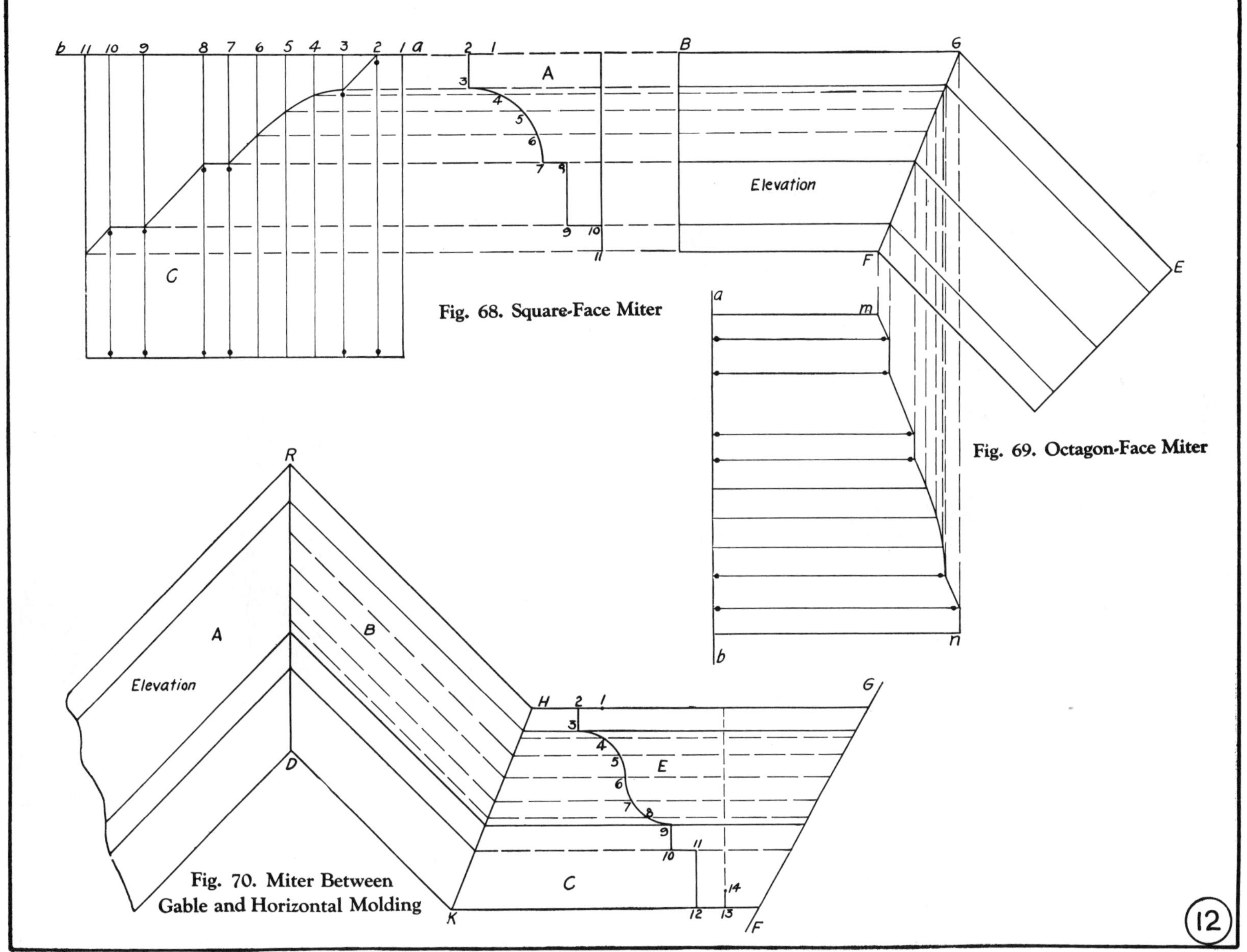

Fig. 68. Square-Face Miter

Fig. 69. Octagon-Face Miter

Fig. 70. Miter Between Gable and Horizontal Molding

From these points, draw lines parallel to *BC*, intersecting the miter line *Cn*. From the points on the miter line, draw lines parallel to *CG*, intersecting the miter line *Ge*. At right angles to *GC* in the elevation, draw the stretchout line *ab*, and develop

Problem 34. Panel Miter.

the pattern for the end of the panel, shown at *H*. This is the only pattern necessary for the construction of the panel, as the miter cut on the end pattern is also used for the long sides of the panel.

Problem 35. Roof Finial

Figure 72 shows the elevation and pattern for a square roof finial, which is drawn to a scale of 3 inches to the foot. Draw the center line *AB* and construct the elevation shown at *C*. Divide the profile into a number of equal spaces, and number the points in the usual manner.

The finial being square in form, the miter on the corner is simply a square return miter and is developed by the short method

shown in Figure 57. Place the stretchout of the profile *C* upon the center line *AB*, which is extended below the elevation. Draw the usual measuring lines, which are intersected by vertical lines drawn from points on the profile in the elevation. Now, measuring from the center line, transfer these points to the opposite side of the pattern by means of the dividers.

A line traced thru the points thus obtained will complete the pattern for one side of the finial, shown at *G*.

Problem 36. Square Ventilator

In Figure 73 is shown the half elevation, half sectional view, and plan view of a square ventilator, which is drawn to a scale of 3 inches to the foot. Let *G* represent the hood of the ventilator and *F* the flange which is joined to the square base, shown at *C*. The band iron brace used in connecting the hood and base is shown at *R*. The half sectional view shows the profile of the different sections, and in actual shop practice, is all that is required for the development of the patterns.

Draw the full size elevation, omit the plan view, and develop the patterns for the different sections by the short method shown in Figure 72.

Problem 35. Roof Finial.

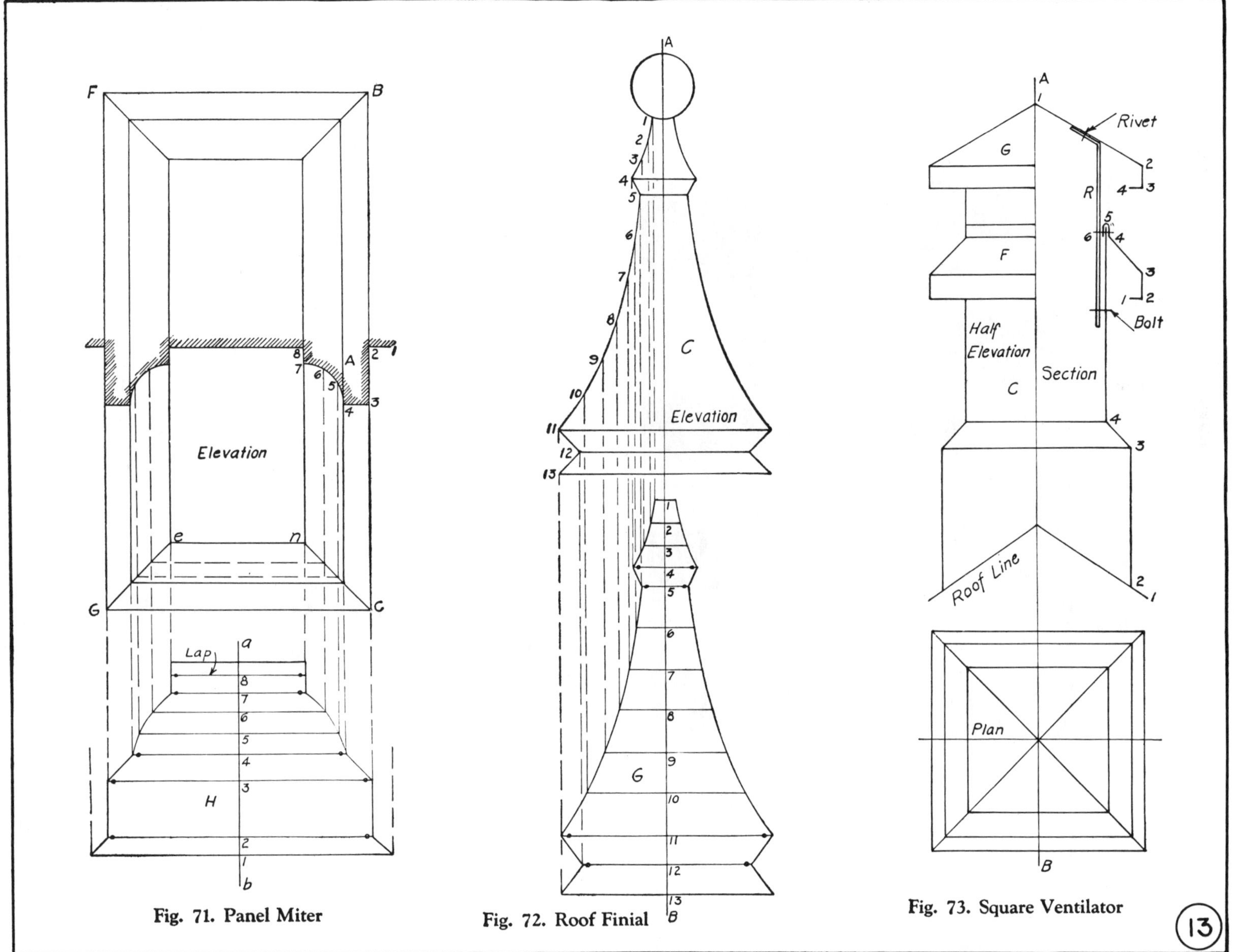

Fig. 71. Panel Miter

Fig. 72. Roof Finial

Fig. 73. Square Ventilator

13

Problem 37. Octagonal Roof Finial

In Figure 74, *A* is the elevation of an octagonal finial, the plan of which is shown at *B*. It is drawn one-fourth size. The principles explained in the previous problems are also applicable to regular polygons having any number of sides. Each angle of the octagonal figure in the plan is bisected, and the bisectors produced until they meet in the center of the octagon. The elevation shows one of the sections, or sides, in profile, and the plan is placed to correspond with the elevation. Divide the profile into equal spaces, and from these points extend vertical lines intersecting the miter lines *ab* and *ac* in plan. From the center of plan at point *a*, and at right angles to *bc*, draw the stretchout line *mn* thru the points, in which draw the usual measuring lines. From the points upon the miter lines *ab* and *ac*, draw horizon-

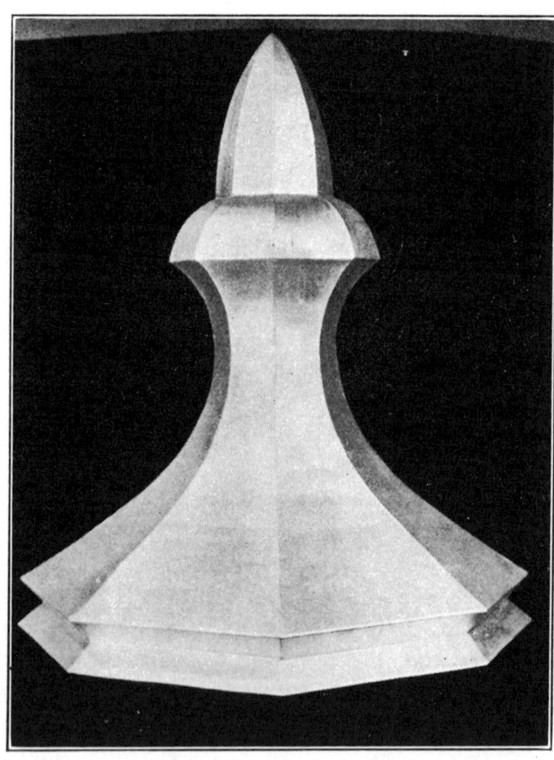

Problem 37. Octagonal Roof Finial.

tal lines intersecting corresponding measuring lines. A line traced thru these points of intersection will describe the pattern for one of the sides of the finial, shown at *C*. The pattern for one section *abc* is all that is required. When developing patterns for any article, the bases of which are regular polygons, the stretchout line must always be drawn at right angles to one of the sides in plan as *bc*, and not at right angles to the miter line *ab*. In actual shop practice, the outlines of the elevation are all that are required for the development of the pattern. Should it be necessary to draw a finished elevation, showing the miter lines in the central portion of the view in Figure 74, the miter lines *1–e* and *1–g* are found in the following manner: Draw lines parallel to *ch* in plan from points on the miter line *ac* to corresponding positions on the line *ah* and *ak*. From these points, vertical lines are drawn, intersecting horizontal lines drawn from similarly numbered points on the profile in the elevation. The foreshortened miter lines *1–g* and *1–e* are then drawn thru intersections of the vertical and horizontal lines.

Problem 38. Conductor Head

Figure 75 shows the front and side elevation of an ornamental conductor head, which is drawn one-fourth full size. The miters on the outer corners are merely square return miters. Draw the front and side elevation and develop the pattern for the front of the conductor head by the method described in Problem 35. The center line divides the pattern into two equal parts. One of these parts will be the pattern for both side pieces, for, in this case, the side of the conductor head is equal to one-half the width of the front. The front elevation is the pattern for the flat back of the head, which is turned in at the top, and the allowance is shown by the dotted lines above the elevation.

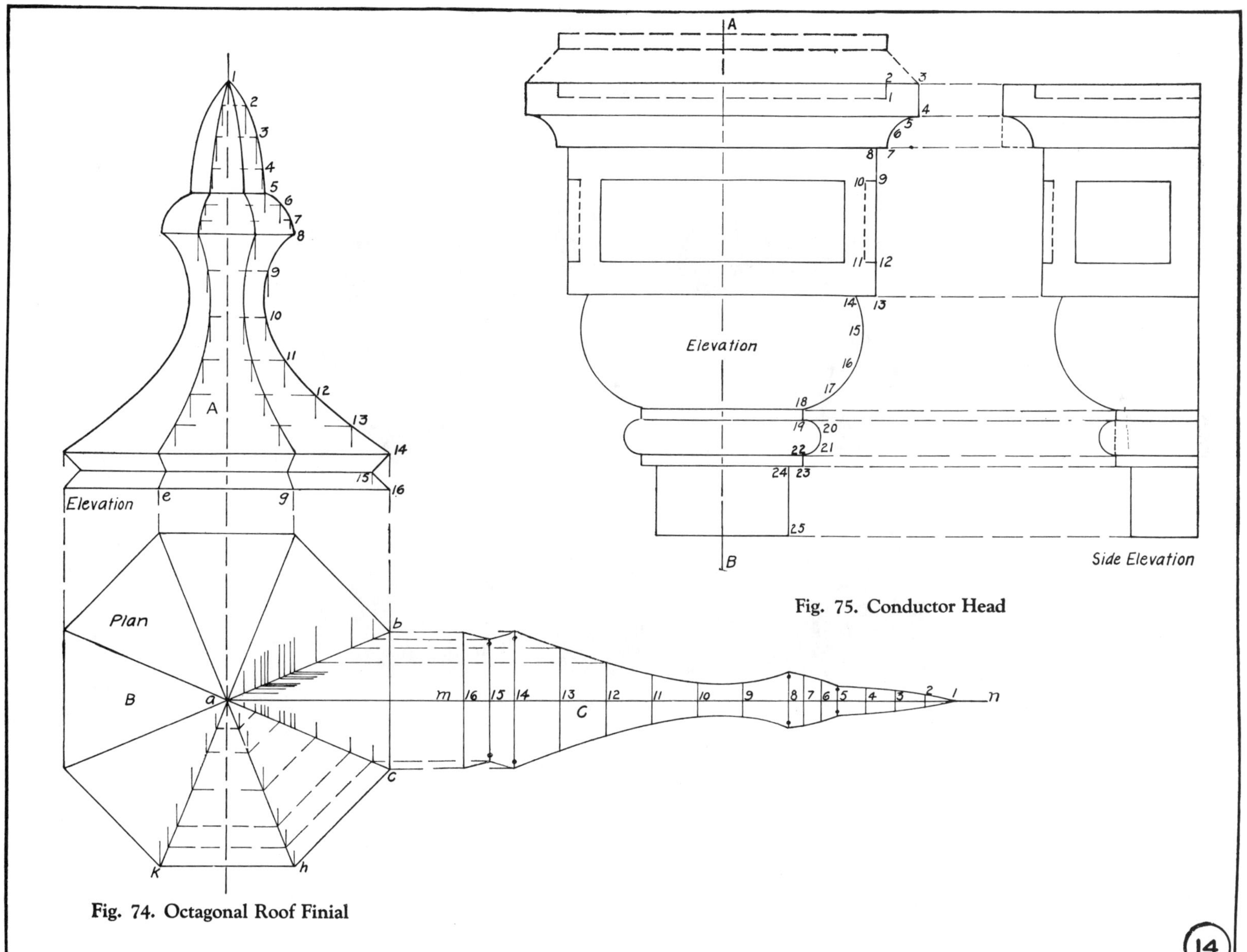

Fig. 74. Octagonal Roof Finial

Fig. 75. Conductor Head

Problem 39. Pediment Molding Mitering on a Wash

Figure 76 shows the half front elevation and side elevation of a pediment mitering upon a horizontal molding, the roof of which is inclined or has what is known as a wash. The drawing is made 4 inches to the foot. This wash, or sloping roof, allows rain and snow to drip off, and should always be placed when the pediment molding has great projection.

Draw the center line *AB*, and on this line place the width of each member of the horizontal molding, as shown by *FEDCB*; also the distance to the top of the pediment, shown by *G*. At right angles to *AB* draw the line *FH* equal to one-half the width of the pediment. From *H* erect a vertical line intersecting the line *C*, or bottom of the wash at point *3*.

Draw a line from *3* to *G* on the center line, which gives the pitch of the pediment.

Draw the line *2–10* at right angles to the pediment mold, and construct the profile as shown at *K*. Divide the cove into equal spaces, shown from *4* to *7*, thru which lines are drawn parallel to the lines of the molding, intersecting the center line *AB* from *1* to *10*, as shown.

Draw the side elevation and construct the profile of the horizontal molding, shown by *abcdefg*. The line of the wash is shown by *cb*. From *c* and *b* draw vertical lines, as shown, and intersect them by a horizontal line drawn from point *G*, which completes the side elevation of the top of the pediment.

Before the pattern for the inclined molding can be described, it will be necessary to obtain the elevation of the miter between the inclined mold and the wash. To obtain this miter, a duplicate of the profile at *K* is placed in the side elevation in the position shown at *P*. From the points on profile *P*, draw vertical lines intersecting the wash *cb*, as shown. Now, from the points of intersection on *cb*, horizontal lines are drawn to the front elevation, which intersect lines drawn from corresponding points on the profile *K*. A line traced thru the points of intersection, as shown from *a* to *3*, will complete the elevation of the miter between the wash and the inclined molding.

The pattern for the inclined molding *K* is shown at *R*, and is developed in the usual manner.

Draw the stretchout line *mn* at right angles to the pitch of the pediment mold, upon which place the girth of profile *K*, as shown by the numbers *1* to *10*. Thru these points draw the usual measuring lines, which intersect by lines drawn at right angles to *G–3* from similarly numbered points on the miter lines, at the top and bottom of the oblique mold.

A line traced thru these points will give the required pattern,

Problem 39. Pediment Molding Mitering on a Wash.

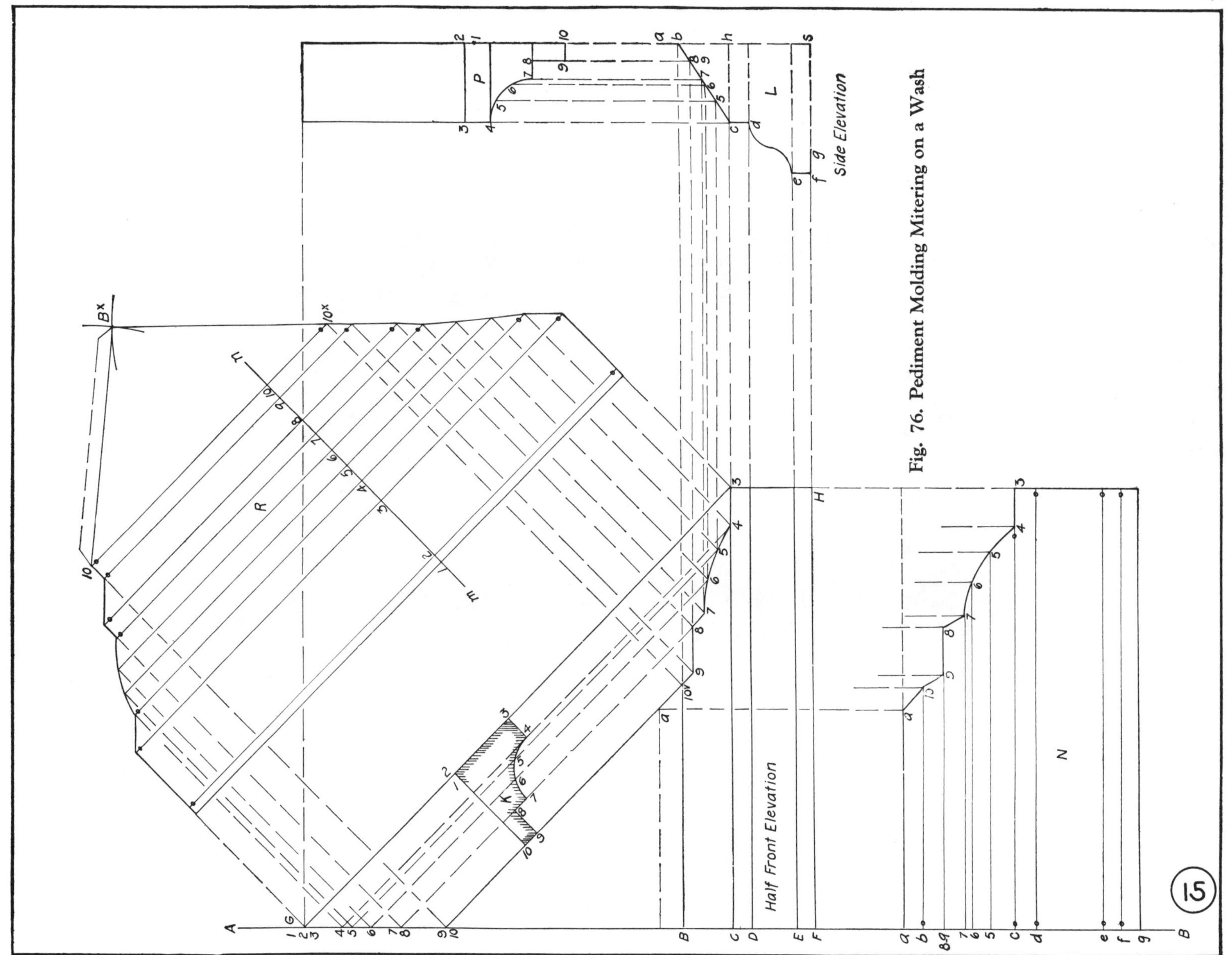
Side Elevation
Half Front Elevation
Fig. 76. Pediment Molding Mitering on a Wash
15

to which is added the triangular piece shown by 10–B^x–10^x in the elevation.   Using B–10 and B–10^v as radii, with 10 and 10^x in the pattern as centers, describe arcs intersecting at B^x.   Draw lines from 10 to B^x, to 10^x, completing the pattern.

The half pattern of the horizontal mold with the miter cut in the wash is shown at N.   To develop the pattern, place the stretchout of the horizontal molding on the center line AB, as shown.   From these points draw horizontal lines, which intersect by vertical lines drawn from similarly numbered points on the miter line in the elevation.

A line traced thru the points thus obtained will be the pattern for the horizontal molding.

The pattern for the two end pieces of the horizontal molding are obtained by taking a duplicate of the profile L, shown by *chsf*, to which laps are added on all sides.

tion of a gable finish or pediment, mitering with a horizontal return molding at the bottom, at a right angle in plan.   In this case, the profile of the return at the bottom is the original, or normal, shown at A, while that of the inclined molding is raked, or changed to correspond, as shown at B.   First, draw the normal profile A and divide it into equal spaces in the usual manner.

From these points draw the lines of the horizontal molding and the inclined molding, as shown.

To obtain the raked profile B, draw the horizontal line bc below the normal profile A.   From the points on profile A, draw vertical lines intersecting the line bc, and number the divisions, as shown, from 3 to 9.   Next, take the various divisions on bc, and place them on the line $b'c'$, which is drawn parallel with the lines of the oblique molding.   From the various intersections and at right angles to $b'c'$, draw lines intersecting similarly numbered lines drawn from corresponding points in profile A.   A line traced thru the points of intersection thus found will give the profile of the raked molding.   The lower part of profile B that miters on the wash is next drawn, and the divisions located upon the line $b'c'$, as shown.

The top of the raked molding extends back to the wall, and the projection is shown by the line 2–3.   The stretchout line mn is next drawn at right angles to the rake molding, upon which

Problem 40.   Gable Molding, with Raked Profile.

Problem 40.   Gable Molding with Raked Profile

The gable ends of buildings are frequently finished with moldings similar to the upper members of the main cornice.   A molding so used is called a raked molding.   When an inclined molding is to miter with a horizontal return, a change of profile in one or the other of the moldings is required, to insure a correct miter of the moldings.   In Figure 77 is shown the front and side eleva-

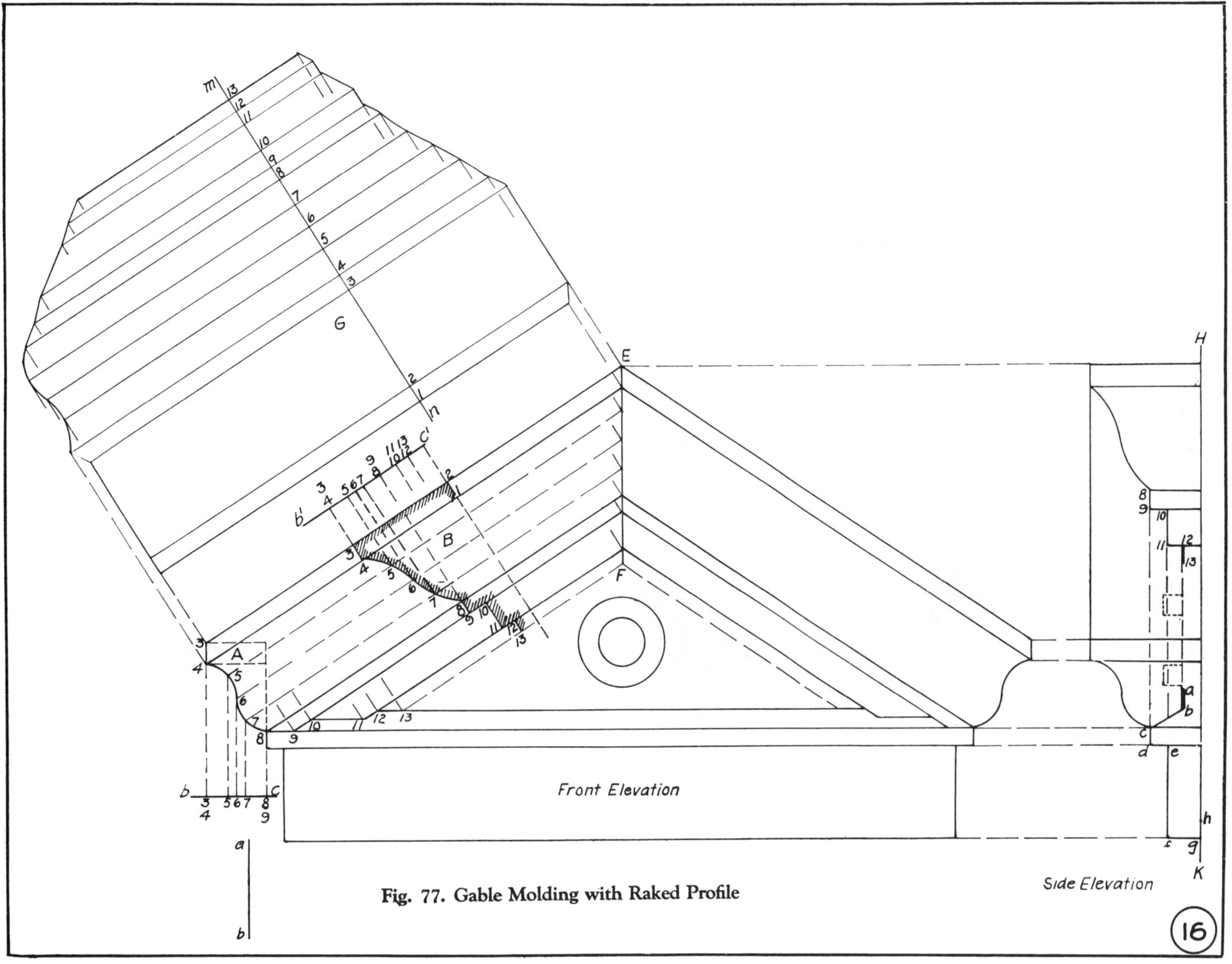

Fig. 77. Gable Molding with Raked Profile

place the stretchout of profile *B*, as shown from *1* to *13*. Each space of the curved portion of the profile must be taken separately, for the divisions are of unequal length.

Next, draw the usual measuring lines which are intersected by lines drawn from points on the normal profile *A*, and the miter line *EF*, thus completing the pattern, shown at *G*.

The pattern for the return at the bottom of the raked molding is a square return miter, which miters upon both the raked and horizontal moldings, and is developed from the profile *A* in the usual manner. The elevation of the miter between the raked molding and the wash—also the

Problem 41. Broken Pediment.

pattern for the horizontal molding—are obtained in the same manner as described in the previous problem. The operations are shown in Figure 76. The profile of the horizontal molding is shown by *abcdefgh* in the side elevation. Observe that the inclined wash shown by the line *cb* is not extended to the wall line *HK*.

Problem 41. Broken Pediment

A pediment is an ornamental form of gable finish for door and window openings, and is frequently placed over the top of the main cornice of a building.

The half front and side elevation of a form of this kind, known as the broken pediment, is shown in Figure 78, which is drawn to a scale of 4 inches to the foot. The drawing shows a section of the horizontal molding and one arm of the inclined molding, which is cut off some distance below the center, and does not extend to the upper miter as in Figure 77. The open space at the top of a broken pediment is usually filled in with some ornamental form.

This problem shows the method used to rake the return moldings when the normal profile is used for the inclined molding, and the process is exactly opposite that shown in Figure 77.

First, draw the line *GH* to the required angle of the inclined molding. The normal profile is then drawn in the position shown at *A*, and is divided into equal spaces in the usual manner. From these points on the profile and at right angles to *GH*, draw lines intersecting the line *mn*, which is drawn parallel to *GH*. Next, take the various divisions from *3* to *10* on the line *mn* and place them on the horizontal line *m'n'*, as shown.

From the divisions on *m'n'* draw vertical lines intersecting similarly numbered oblique lines drawn from corresponding points in profile *A*. A line traced thru the points of intersection will give the raked profile *B*.

To find the raked profile *C* at the upper end of the inclined molding, establish the point *3'* on the line *GH* and take the various divisions from *3* to *13* on *mn* and place them on the horizontal line m^2n^2, having the point *3* placed directly over *3'*. From the divisions on the line m^2n^2, draw vertical lines, intersecting similar lines in the inclined molding. The raked profile *C* for the upper return molding is then traced thru the points of intersection.

The pattern for the inclined molding is shown at *F*, and the

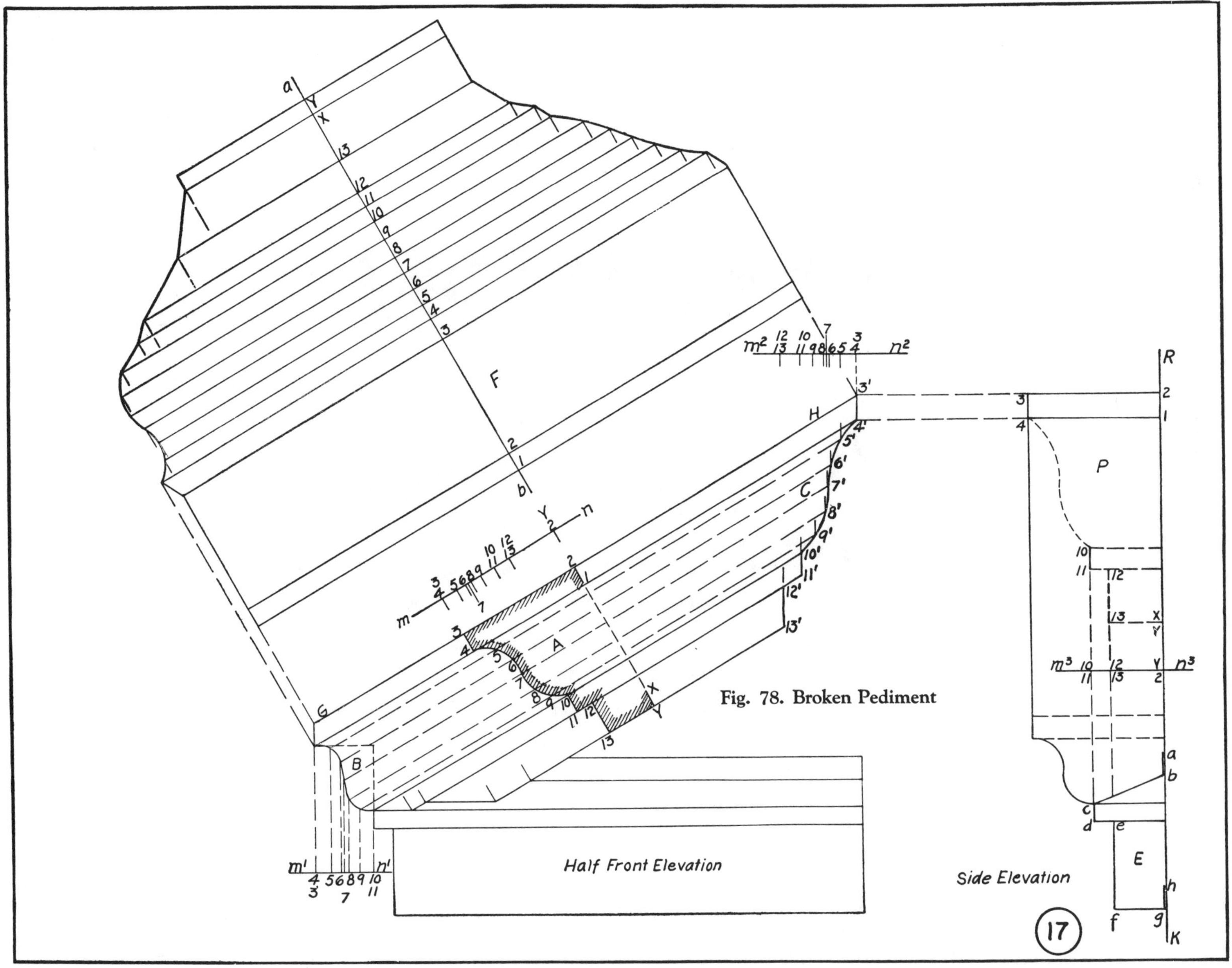
Fig. 78. Broken Pediment
Half Front Elevation
Side Elevation
17

method of development is similar to that described in Problem 40.

The pattern for the upper and lower returns are simply square return miters, and further description is unnecessary.

The profile of the horizontal molding is shown at E in the side elevation, the slope of the wash is represented by the line cb, and extends back to the wall line RK.

The method of finding the miter between the inclined and horizontal moldings—also the cut on the wash of the horizontal molding—is fully shown in Figure 76. In actual shop practice, it is unnecessary to draw a side elevation of the inclined molding shown at P to obtain the points 12–13 on the wash cb. The required points can be found in the following manner: At any convenient distance above the line cb, representing the wash, and at right angles to the wall line RK, draw the line m^3n^3. Take the divisions 10–11, 12–13 and Y–2 on the line mn, and place them on the line m^3n^3, as shown, being careful to place the point Y–2 on the wall line RK. Vertical lines drawn from these points to the line cb will give the required points on the wash without drawing the side elevation of the inclined molding.

Problem 42. Miter Between Moldings of Different Profiles

It is often necessary in cornice work to bring moldings of different profiles together in a miter. To develop patterns for a square return miter between moldings of dissimilar profiles requires two distinct operations, and the miter cut upon each piece would be the same as it would appear when a butt miter was made between the two moldings. Figure 79 shows the pro-

files and patterns for a square return miter between moldings of two different profiles, which are drawn one-third full size. The two arms of the miter being different in profile, it is necessary to draw an elevation of the two moldings in the position shown at A and B. The profiles having been drawn in the position shown, the patterns are developed in the same manner as the butt miter, shown in Figure 59. Divide the profile A into any convenient number of parts, from which draw horizontal lines intersecting profile B, as shown. To develop the pattern for molding A, place the stretchout of the molding upon the line ab, which is drawn at right angles to the lines of the molding.

Draw the usual measuring lines which are intersected by vertical lines drawn from similarly numbered points on profile B.

A line traced thru these points will give the miter cut of the pattern for molding A, to fit against the profile B.

The pattern for an outside miter is shown at C, and an inside miter at E. To obtain the pattern for molding B, proceed in the same manner, reversing the order of the profiles. Develop the stretchout of molding B, on the line mn, being careful to take each division separately, as they are of unequaled width. Draw the horizontal measuring lines which are intersected by vertical lines drawn from similarly numbered points on profile A.

A line traced thru these points will give the pattern for molding B, which miters with profile A. An outside miter pattern is shown at F, while the opposite cut is an inside miter, shown at G.

Problem 42.   Miter Between Moldings of Different Profiles.

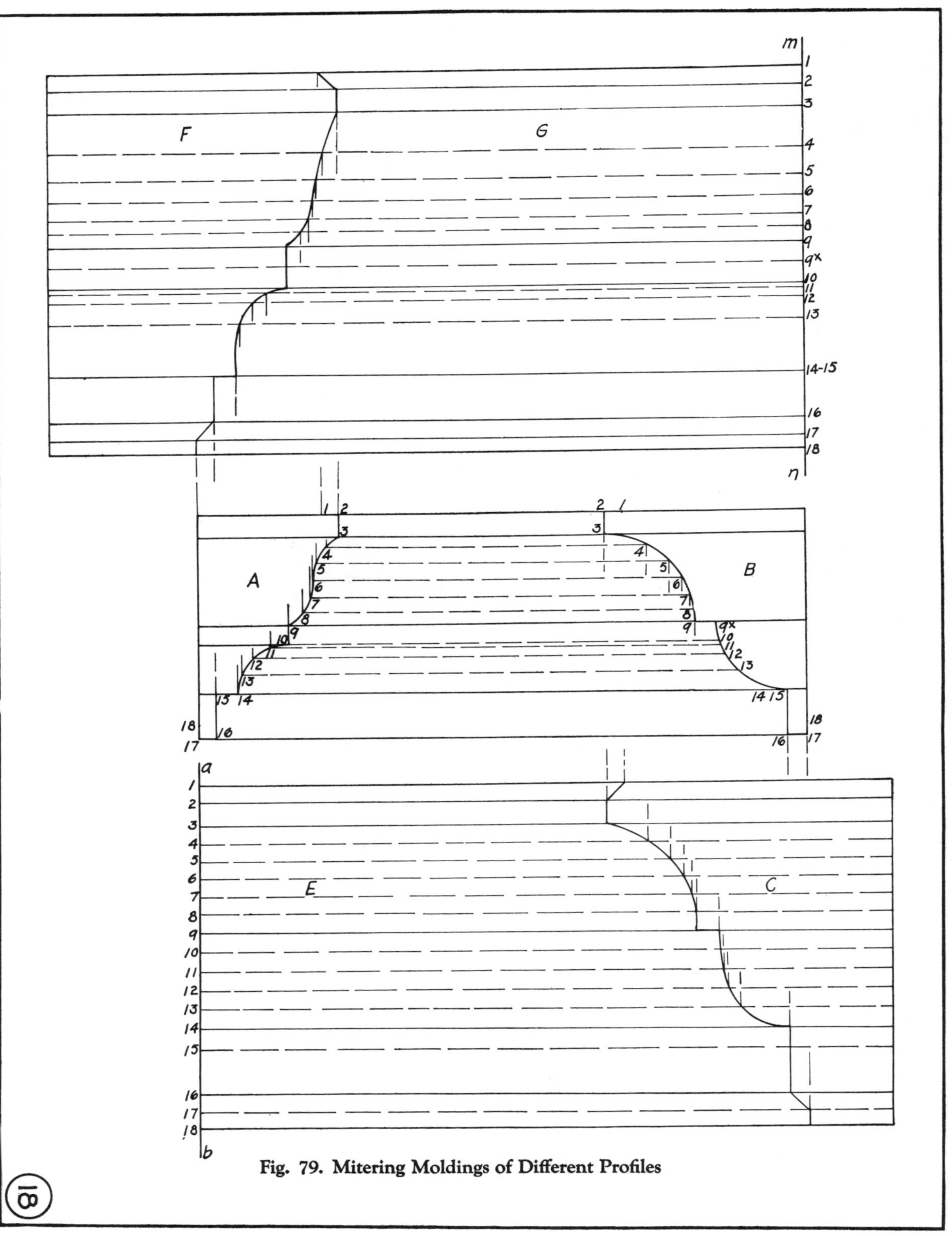

Fig. 79. Mitering Moldings of Different Profiles

PART THREE
RADIAL DEVELOPMENTS

CHAPTER VI
REGULAR TAPERING FORMS

This subject embraces a large variety of forms of frequent occurrence in sheet-metal work, and the forms or shapes considered within this part include only such forms that have for their base the circle or any of the regular polygons, as the square, hexagon, octagon, etc.; also figures, tho of unequal sides, that can be described within a circle, in which lines drawn from the corners terminate in an apex over the center of the base.

When developing patterns for tapering forms, the following rules will enable the student to understand the principle by which these developments may be accomplished.

1) A drawing must first be made consisting of an elevation showing the true height of the apex, and the true length of the radius with which to describe the stretchout of the pattern.

2) A plan view must be drawn from which the length of the stretchout can be obtained, as shown by EF in Figure 80.

3) The stretchout arc must be described with a radius equal to the length of the true edge of the solid, as shown by AH Figure 80.

4) Points are located on the stretchout corresponding to the position of the points on the outline of the plan or sectional view.

5) Measuring lines and edge lines of the pattern are always radii of the stretchout arc.

6) Should an irregular or straight line be drawn through any cone, as illustrated by the line G-9 in Figure 82, and which the radial lines in elevation will intersect, then, from these points of intersection on G-9, lines must be drawn at right angles to the axis AH intersecting the side of the cone AB, which will give the true lengths from apex A, and are carried to similarly numbered radial lines in the pattern R, as shown from 1 to 1.

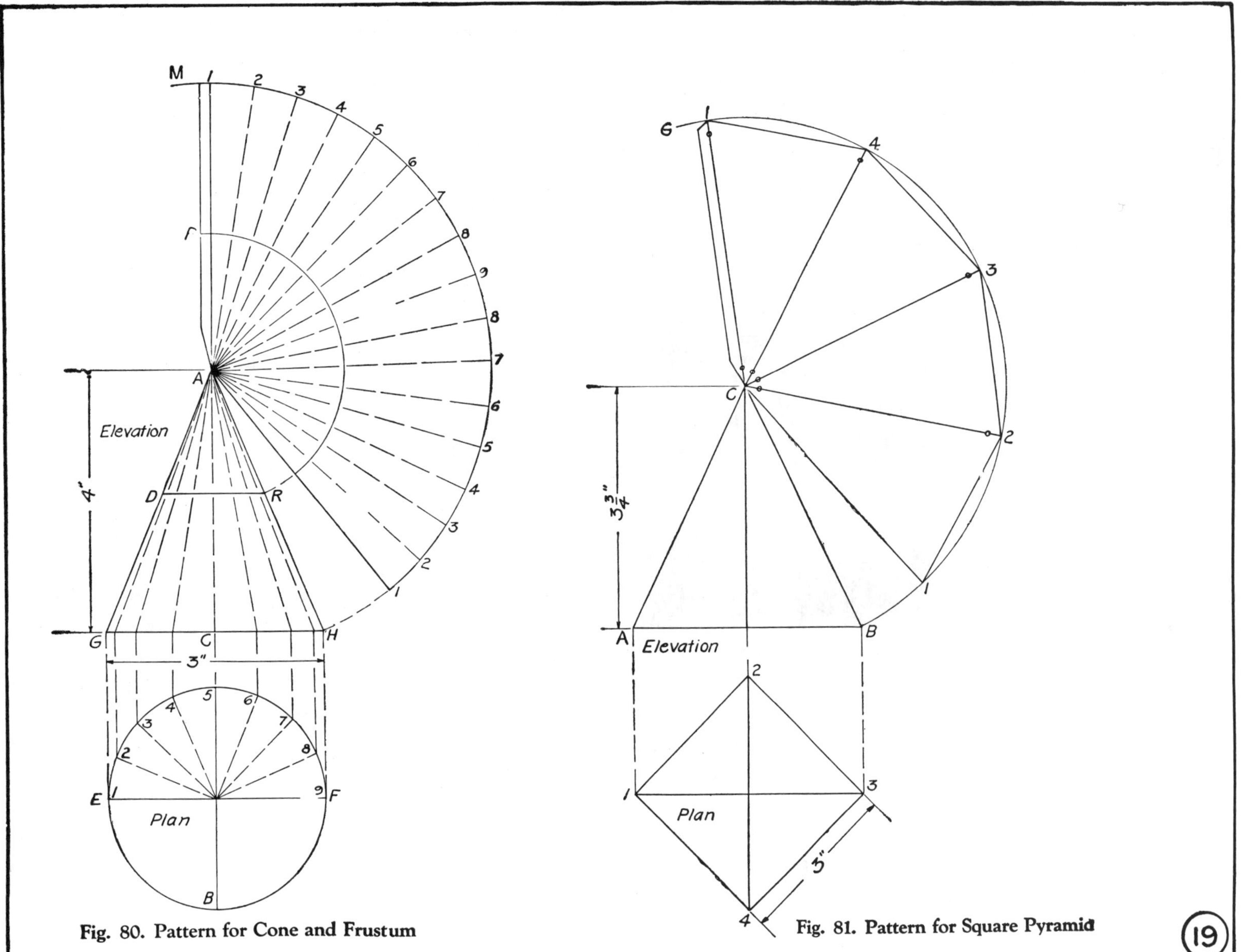

Fig. 80. Pattern for Cone and Frustum

Fig. 81. Pattern for Square Pyramid

19

Problem 43. Pattern for Cone and Frustum

In Figure 80 is shown the method of developing the pattern for a right cone. This method contains the principles applicable to all pyramids which have for their base the circle or any of the regular polygons that can be inscribed within a circle.

First draw the center line AB, upon which place the height of the cone as AC; thru C draw the horizontal line GH equal to the width of the base, and draw lines from G and H to A. Directly below the elevation, describe a circle to represent a plan view of the base as shown by EF, which is divided into equal spaces as shown from 1 to 9.

The radial lines shown in the elevation, plan and pattern are not necessary, but are shown to make clear their relation to each other. To obtain the pattern for the cone, use the apex A as center and with a radius equal to the true length of the slant height of the cone as shown by AH, describe the stretchout arc HM. On any convenient point on the stretchout locate point 1 and draw a line from 1 to A. Then set off on the arc HM of the pattern, commencing at 1, twice the number of spaces that are contained in the half circumference of the plan, as shown by 1–9–1. From the end point thus located draw a line to the apex A, and then add an allowance for seaming. If a frustum of a cone is desired, as shown by $GDRH$, then using AR as radius, draw the arc RP, making R–P–1–1 the desired pattern.

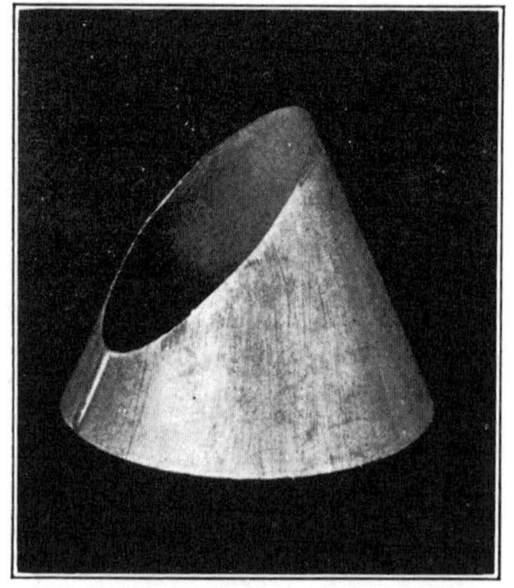

Problem 45. Irregular Frustum of a Cone.

Problem 44. Pattern for a Square Pyramid

This development is shown in Figure 81. Draw the plan and elevation according to the dimensions given in the drawing. When the plan view is placed in the position as shown, the line CB in the elevation represents the true length of one of the corners of the pyramid; the stretchout arc may, therefore, be described as in the case of the cone in the preceding problem. With C as center and CB as radius, describe the arc BG. After setting the dividers to the width of one side of the base shown in the plan at 1–2, begin at 1 and step off on the stretchout arc BG spaces equal in number to the sides of the pyramid. Thus, points are located at 1, 2, 3, 4 and 1. Connect these points by straight lines as shown, and draw lines from each point to the apex A, completing the drawing. The outline C–1–2–3–4–1 is the development of the pyramid.

Problem 45. Pattern for an Irregular Frustum of a Cone

Figure 82 shows the method of developing the pattern for the frustum of a right cone cut by the plane represented by the line G–9 drawn oblique to the axis of the cone, which also intersects the radial lines from 1 to 9.

First draw the elevation of the cone ACB, and directly below it the plan view D. Divide one-half of the plan into equal spaces as shown by the figures 1 to 9. Next draw the line G–9 at an angle of 45° with the base line CB, making the point G one inch from C. From the various points in the plan erect lines intersecting the base of the cone from 1 to 9, and from these intersections draw lines to the apex A, intersecting the line G–9 as shown.

Using A as center and with AB as radius, describe the stretchout arc BF, upon which step off twice the number of spaces

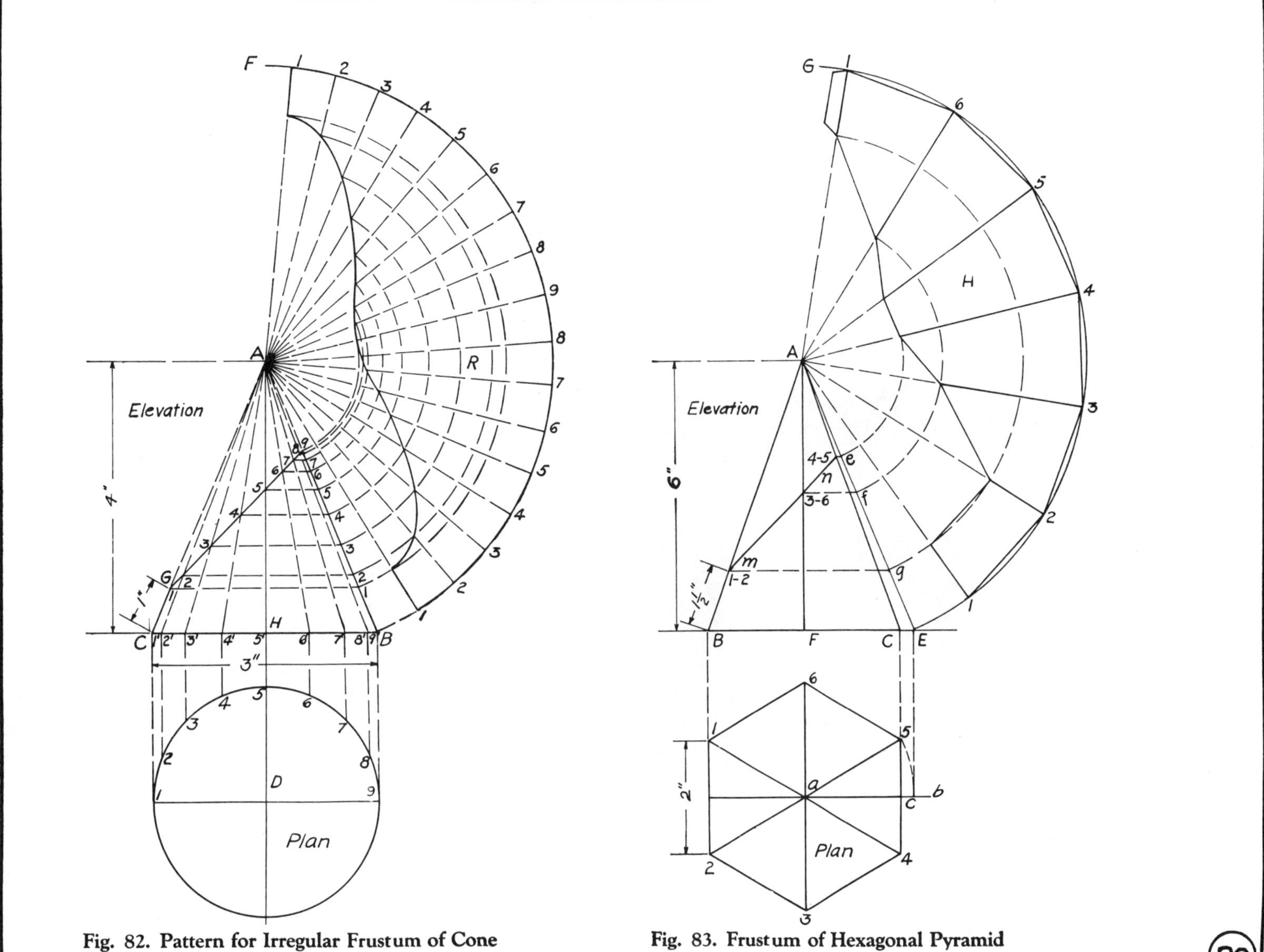

Fig. 82. Pattern for Irregular Frustum of Cone   Fig. 83. Frustum of Hexagonal Pyramid

shown in the plan D, as shown by similar numbers *1–9–7*. From these points draw radial lines to the apex A. From the various intersections on the line G–*9*, and at right angles to the axis line AH, draw lines as shown intersecting the side of the cone AB from *1* to *9*. Then, using A as center with radii equal to the various divisions as A–*1*, A–*2*, A–*3*, etc., draw arcs intersecting similarly numbered radial lines in the pattern R. The irregular curve traced thru points thus obtained completes the desired development.

Problem 46.   Pattern for an Irregular Frustum of a Hexagonal Pyramid

This development is shown in Figure 83, the plan and elevation being first drawn according to the dimensions given in the figure. The cutting plane is shown by the oblique line mn and is drawn at an angle of 45° with the base line.

The true length of the edge lines in the pyramid are not shown in either the plan or elevation, and it is, therefore, necessary to draw a line that will represent the true edge in the elevation. This is found

Problem 46.   Irregular Frustum of a Hexagonal Pyramid.

as follows: From a as center of the plan view, with the radius a–*5*, describe the arc *5*–c, intersecting the line ab at c. From c draw a vertical line to the base line BC of the front elevation extended at E. AE is then the true length of one of

the corners of the pyramid, and is also the true length of the radius of the stretchout arc.

The stretchout arc EG is next described from A as center with a radius AE. The widths of the sides of the base are then laid off at *1–2–3–4*–, etc.; connect these points by straight lines, as shown, and draw lines from each point to the apex A.

From the intersections on the oblique line mn and at right angles to the axis AF, draw lines to the true edge line of the pyramid at efg. With A as center and radii Ae, Af and Ag, describe arcs intersecting similarly numbered radial lines in the pattern, shown at H.

Complete the development by drawing lines connecting these points in the manner shown.

Problem 47.   Conical Gutter Outlet

In many cases of eave-trough construction, where it is desired to so connect the conductor pipe that the opening in the gutter will be larger than the diameter of the pipe, a tapering connection pipe is used, as shown in Figure 84.

An examination of the drawing shows that the flaring outlet consists of a frustum of a cone; its upper base in the drawing is defined by the straight line FH, and the lower base by a section of the gutter. First, draw the center line AB and construct a section of the gutter, shown by the arc CGD. Make the distance from G to O equal to the height of the outlet, and thru O, at right angles to the center line AB, draw the line FH equal in length to the diameter of the conductor pipe.

Draw the line mn, intersecting the center line at R. Then, using R as center, describe the half plan of cone, as shown by men. From m and n, draw lines thru F and H, intersecting center line at A, completing the elevation of the entire cone.

Develop the patterns for the conical outlet $mFHn$, using

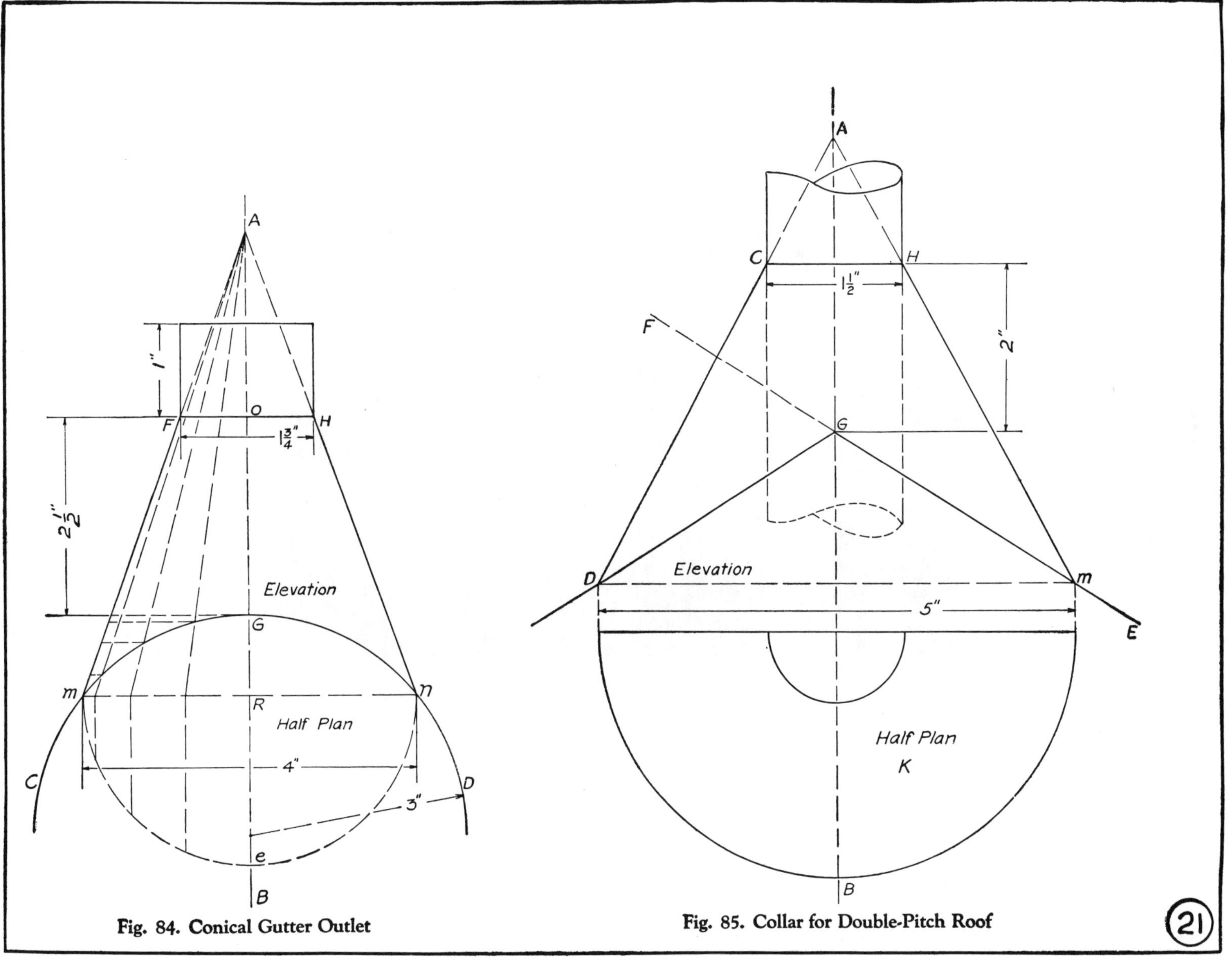

Fig. 84. Conical Gutter Outlet

Fig. 85. Collar for Double-Pitch Roof

21

principles similar to those used in Problem 45, in the development of an irregular frustum of a cone.

Problem 48. Tapering Collar for Roof Having a Double Pitch

The principles used in developing the pattern for the intersected cone in Figure 82 are applicable, no matter at what angle or point the bases of the cone are intersected.

Develop the patterns for a tapering collar, shown by *C HmGD* in Figure 85, the base of which is to fit against a roof of two inclinations, as indicated by *DGE*; also when it has a single pitch, as shown by *FE*. The dotted line *Dm* represents the base, and *A* the apex of the cone. The half plan is shown at *K*, below the elevation.

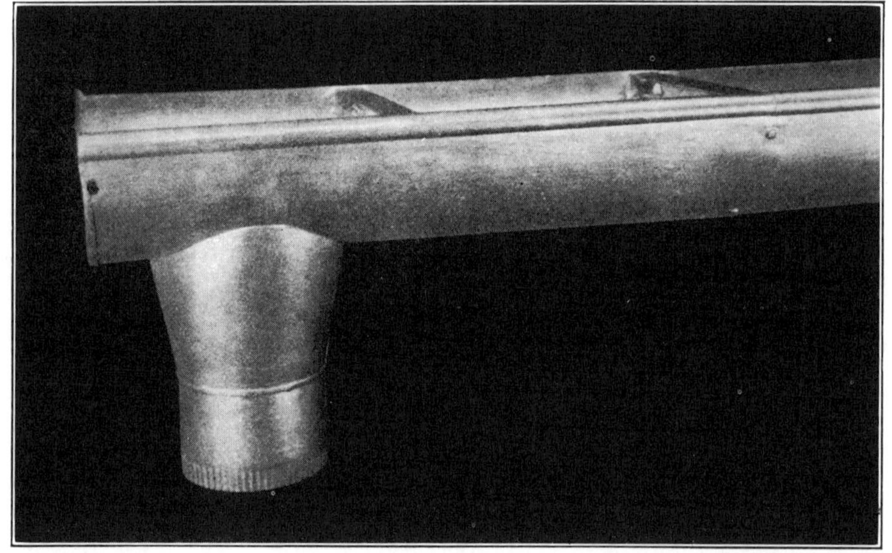

Problem 47.   Conical Gutter Outlet.

upon the metal by a method in which no elevation is required, as the true length of the radius for describing the stretchout arc is found in the plan. The length, width and height of pyramid being known, first draw a plan to the required size, as shown by *1–2–3–4*, and then the diagonal lines *1–3* and *2–4*, which represent the miter lines or hips, intersecting in the center at *a*. Bisect the line *1–4* and locate the point *e*; then draw a line from *e* to *a*, showing the position of the seam.

Before describing the stretchout arc for the pattern, the true length of one of the hip lines in plan must be found, and that dimension used as the radius for describing the stretchout arc.

To find the true length of the hip line *a–2* in plan, erect the perpendicular *a–b* equal to the vertical height *AM* in the elevation. Draw a line from *2* to *b*; then *2–b* is the true length of the line *2–a* in the plan, and is the radius with which to develop the pattern. With any point with *F* as center, describe the circle *g–d*.

Starting at any convenient point on the arc, as point *2*, step off the length of the end and sides of the pyramid, shown by the divisions *1–2–3–4*, and draw lines to the apex *F*. From *1* and *4* as centers with a radius equal to one-half the width of the base, as shown by *1–e* in plan, describe the short arcs, as shown. The

Problem 49.   Rectangular Pyramid

Figure 86 shows the development of a pattern for a rectangular pyramid. This principle is applicable to various-shaped ornaments in cornice work; also all pyramids that have for their base any of the regular polygons that can be inscribed within a circle, the lines drawn from the corners of which would terminate in an apex directly over the center of its base.

Patterns for work of this kind are usually laid out directly

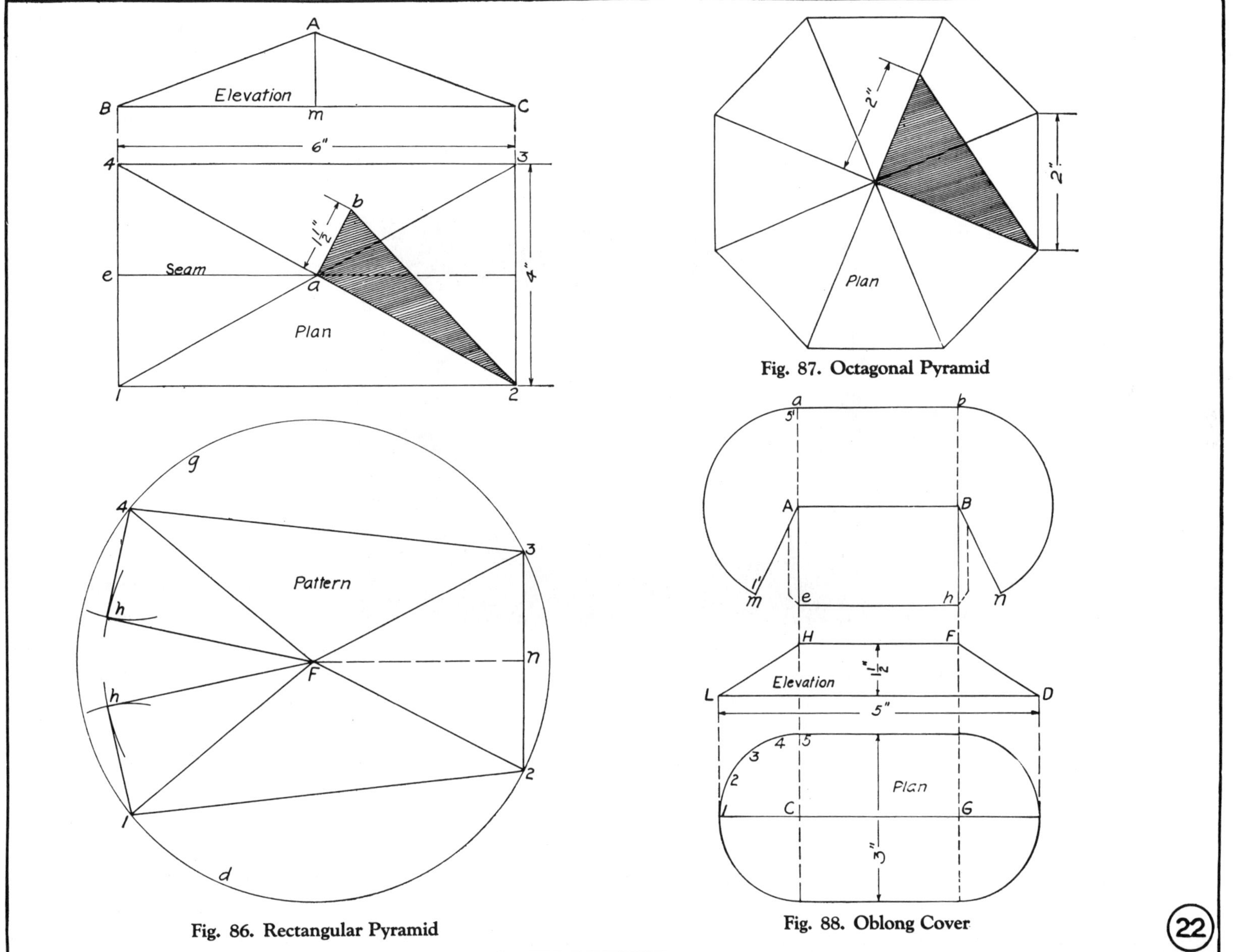

Fig. 86. Rectangular Pyramid

Fig. 87. Octagonal Pyramid

Fig. 88. Oblong Cover

22

true length of the seam line is shown by the dotted line *Fn*. Then with *Fn* as radius and *F* as center, describe arcs at *hh*. Connect all points by straight lines, completing the pattern.

Problem 50.   Octagonal Pyramid

From Figure 87, draw the plan view and develop the pattern for an octagonal pyramid. Apply the method used in Problem 49.

Problem 51.   Oblong Pitched Cover

Figure 88 shows the method of developing the pattern for an oblong raised cover with semi-circular ends. First draw the plan and elevation, which will show that the shape consists of the two halves of the envelope of a right cone, connected by a straight piece in the center. To develop the pattern, proceed as follows: Draw the line *AB* equal in length to *CG* of the plan. From *A* and *B* as centers, with radius equal to *LH* of the elevation, describe the stretch-out arcs, as shown by *am* and *bn*. Upon these arcs, from *a* and *b*, step off twice the number of spaces which are contained in one-half the semi-circular end, as shown by the divisions *1* to *5* in plan, thus obtaining the points *m* and *n*. From *m* draw the line *ma*, and from *n* draw *nB*. From *A* and *B*, at right angles to the center line *AB*, draw the line *Ae* and *Bh* equal in length to *Am* of the pattern.

Connect *e* and *h*, completing the development.

Problem 52.   Flaring Pan

In Figure 89 is shown the elevation and half pattern for a flaring pan made in two pieces, the form of which is seen to be the frustum of a right cone. An inspection of the drawing will show that *AB*, the top of the pan, is the base of an inverted cone, its apex *C* being at the intersection of the lines *AG* and *BH* forming the sides of the pan, and that *GH* is the top of the frustum or the base of another cone which remains after taking the frustum from the original cone. In developing the pattern, first draw the center line *FC*, upon which place the height of the pan *FR*; thru these points draw lines at right angles to the center line. On either side of the center line from the points *FR*, place the half diameter *FB* of the top and *HR* of the bottom. Draw lines connecting *BH* and *FA* and extend them until they meet the center line in the point *C*. With *R* as center and *RH* as radius, describe the quarter circle *HD*, and divide it into a number of equal parts, as shown by the figures *1* to *7*. This quarter-circle represents a one-quarter plan of the bottom of the pan. The pattern is developed as follows. With *C* as center and the radii equal to *CH* and *CB*, draw the arcs *OE* and *KN*, as shown. From the center *C*, draw a line across these arcs near one end, as *KO*, and starting from the point *K*, step off on the arc *KN* twice the number of spaces contained in the quarter plan, as shown by the figures *1–7–1* on the arc. Thru the last

Problem 52.   Flaring Pan.

Half Elevation
30°
A
3/8"
4"
2"
C
1 1/4"
45°
F
5/8"
5"
B
4"
G
Section
Fig. 91. Round Ventilator Head
A
8 7 6 5 4 3 2
m
Half Elevation
n
C
B
G
Fig. 90. Pattern for Funnel
6 1/4"
A F B
2 5/8"
G R
4 3/4"
H
Quarter Plan
1 2 3
D 7 6 5 4
K
1 2 3 4 5 6 7 6 5 4 3 2
C N E
Fig. 89. Flaring Pan
6"
C
4"
B
1/4"
2"
1 1/8"
G
7/8"
Fig. 92. Fruit Jar Filler

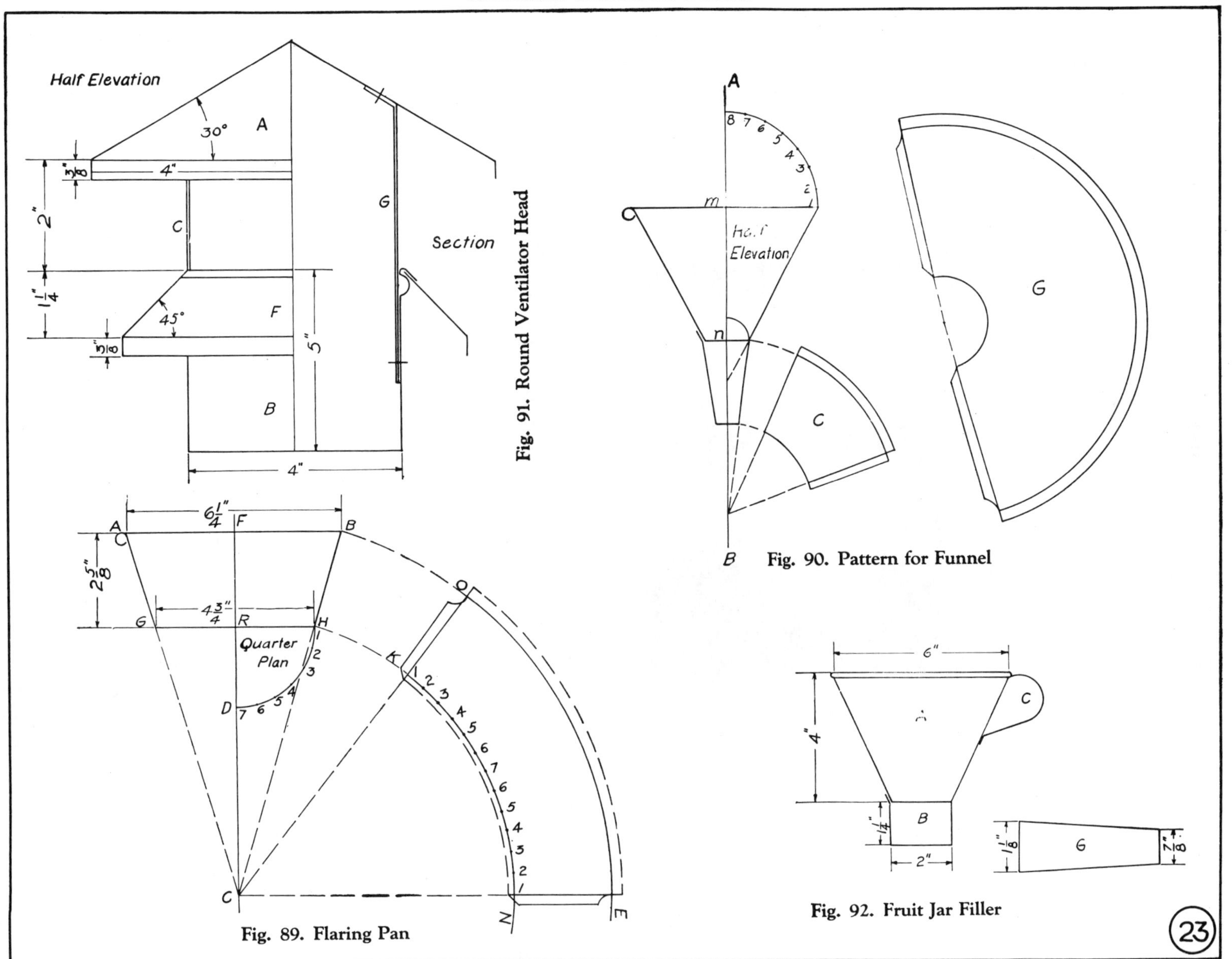

23

division, draw a line across the arcs to the center C, as shown by $E N$. Add laps for wiring and seaming, as shown by the dotted lines. A one-half elevation and a quarter plan of the top or bottom is all that is required to find the stretchout and radius for describing the pattern for the frustum of a cone.

Problem 53.   Pattern for a Funnel

Figure 90. Applying the method given in Figure 88, develop the patterns for the body and spout of the funnel, shown in the drawing. The vertical height is 3-1/2 inches. The diameter of the top is 5 inches, and the lower opening in the body measures 1 inch in diameter. The spout is 2 inches long and has a 1/2-inch outlet, the seam being lapped and soldered. It will be seen from Figure 90 that all the operations required may be performed if only one-half of the elevation be drawn,

Problem 54.   Round Ventilator Head.

as shown on one side of the vertical center line AB.

A one-quarter plan of the top and bottom is shown in the half elevation by the quarter-circles which are described from m and n as centers. As a matter of convenience, the pattern for the body, shown at G, is laid off to one side of the elevation, and the stretchout arc is not here described from the apex of the cone as was conveniently done in the previous problem. The pattern for the spout is shown at C.

Problem 54.   Round Ventilator Head

In Figure 91 is shown the half elevation and half section of a round ventilator head. The conical hood is shown at A, and the lower flange at F. The supports C and G are made from band iron riveted to the hood A and the round pipe B, as shown in the drawing. The lower flange F has an inclination of 45°, and the pitch of the upper hood A is at an angle of 30°. Draw the full elevation and develop the patterns for the upper hood A and the flange F.

Problem 55.   Fruit-Jar Filler

Figure 92. Develop the patterns for a fruit-jar filler. The flaring body is shown at A, the spout at B, and the handle at C. The pattern for the handle is shown at G, to which an allowance has been added for a 3/16-inch hemmed edge on each side.

Problem 56.   Elliptical Flaring Pan

The plan, elevation and half pattern for an oval flaring pan are shown in Figure 93. As may be understood from the drawing, the bases are elliptical, while the sides flare uniformly. The short and long diameters of the lower base or top of the article are 9 x 12 inches, respectively. The sides flare 1-1/2 inches all around; that is, the upper base or bottom is an ellipse, whose long and short diameters are 6 and 9 inches, respectively. The vertical height is 3-1/2 inches. Draw the plan view according to the rule given in Figure 26 (Practical Geometry) and locate the centers a, b, e, m, as shown. Next, draw the elevation $BFGH$, setting off the vertical height 3-1/2 inches. The next step is to obtain the radii with which to strike the arcs of the pattern.

From b and m in the plan, draw the vertical lines bA and mE, and extend the side HF of the elevation until it intersects the perpendiculars from b and m in the points A and E. Then $A H$

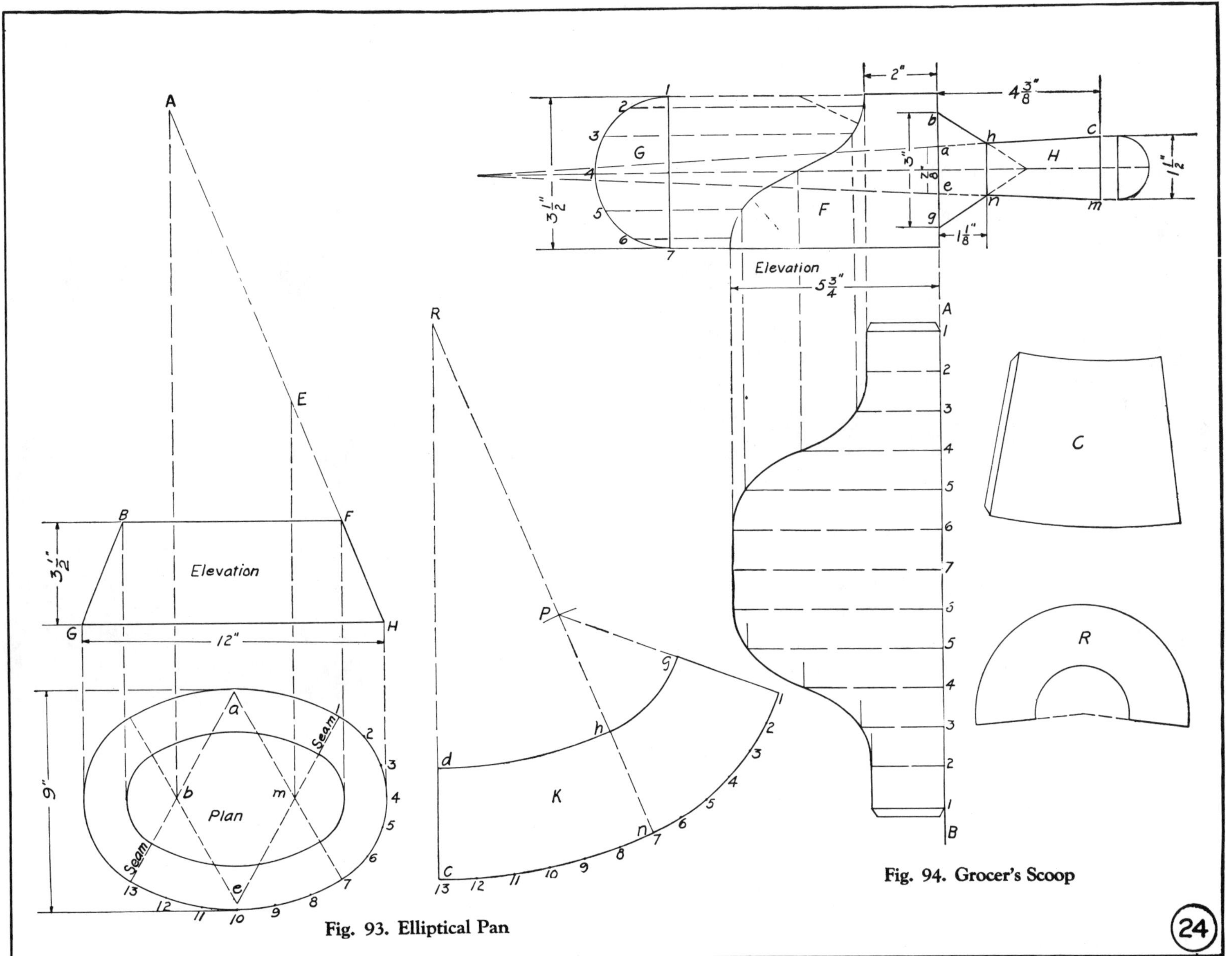

Fig. 93. Elliptical Pan

Fig. 94. Grocer's Scoop

24

is the radius with which to strike the pattern for that part, shown by *13–7–a* in the plan, and *AF* the radius for that part, shown by *7–1–m* in plan.   With *A H* as radius and *R* as center, describe the arc *cn* in pattern *K*.   Starting at any point as *c*,

Problem 56.   Elliptical Flaring Pan.

set off on the arc *cn*, the stretchout of the arc *13–7* in plan. Draw lines from *13–7* to *R*.

Then with *E H* in elevation as radius and *n* in pattern *K* as center, describe an arc cutting the line *nR* in *P*.

With *P* as center and *Pn* as radius, describe the arc *7–1*, upon which set off the stretchout of *7–1* in plan.   Draw a line from *1* to *P*.   Next, with radii equal to *AF* and *EF* in the elevation and with centers *R* and *P*, describe the arcs *dh* and *hg*, thus completing the pattern with seams at *1* and *13*.   Allowances for seaming and wiring must be added to the pattern.

Problem 57.   Grocer's Scoop

In Figure 94 is shown the elevation and patterns for a hand scoop, the body being in the form of an intersected cylinder, and the handle and brace are the frustums of two right cones.

This problem as presented will require the development of

patterns by both the parallel line and radial methods, and the student's attention is directed to the manner in which any irregular section of the cylinder may be produced.

First draw the side elevation and half section to the dimensions given in the drawing.   Divide the half section *G* into a number of equal spaces, shown from *1* to *7*, and from these points draw horizontal lines intersecting the outline of the side elevation.   To obtain the pattern for the body *F*, draw the stretchout line *AB* in the position shown and proceed to develop it in the usual way, the spaces shown by *1–7–1* being set off in the usual manner.

The handle *H* is the frustum of a cone shown by *cmae*, which is soldered to the flat back of the scoop.   The conical boss or brace is shown by *hngb*.

The patterns for the brace and handle are

Problem 58.　Tapering Square Pipe Intersecting a Vertical Square Pipe.

shown at *R* and *C*, and the method of development has been fully described in previous problems.

Problem 58. Tapering Square Pipe Intersecting a Vertical Square Pipe

Figure 95 shows the manner of developing the pattern for a tapering square pipe intersecting a vertical square pipe placed

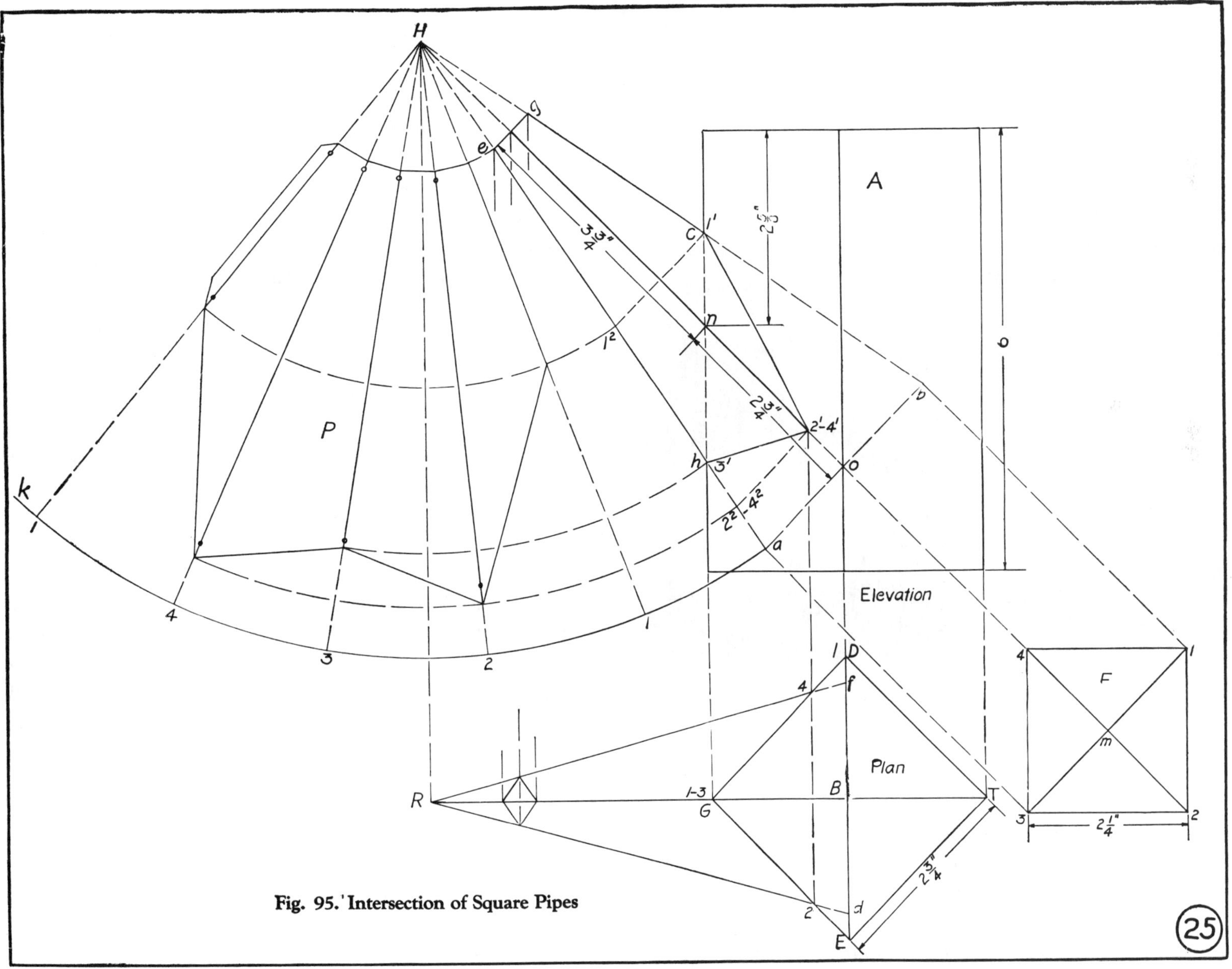

Fig. 95. Intersection of Square Pipes

diagonally in plan.   In this problem, first draw the plan view of the vertical pipe in the position shown by *GDTE*; directly above it, the elevation *A*.   Next, in its proper position in elevation, draw the tapering pipe *egch*, extending the lines of the pipe equally until they intersect in the apex *H* and the base line *ab*.

Below the base line *ab*, draw the section *F*, as shown.   The top and bottom corners of the tapering pipe intersect the corner of the vertical pipe in the plan at *G*, shown in the elevation by *c* and *h*, while the side corners intersect the flat sides of the vertical pipe at *2–4* in the plan and elevation, and this point must be found as follows: From the apex *H* in elevation, drop a vertical line intersecting the center line of plan at *R*.   Next, take the distance *m–2* and *m–4* in section *F*, and from *B* in center of plan, place this dimension upon the line *DE*, as shown at *f* and *d*. Draw the lines *Rd* and *Rf*, cutting the sides of the ver-

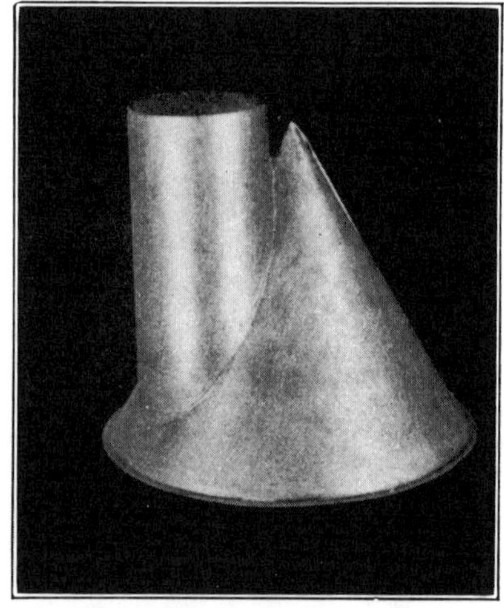

Problem 59.   Cone Intersected by Vertical Cylinder.

tical pipe at *2* or *4*.   From these points, draw a vertical line intersecting the center line *HO* in elevation at *2'–4'*.   Connect *ch* and *2'–4'* by straight lines, which will give the miter or line of intersection between the two pipes.   Before the pattern for the tapering pipe can be developed, the true length of the corners must be found as follows: From the intersections *1'* and *2'–4'* at right angles to the center line *HO*, draw lines until they intersect the line *eh*, the lower corner of the tapering pipe.

To obtain the pattern for the pipe, use *H* as center, and with *Ha* as radius, describe the arc *ak*.   After setting the dividers to the width of one side of the base, shown in section *F* by *1–2*, begin at *1* and step off on the stretchout arc *ak*, spaces equal in number to the sides of the pipe, and from these points draw radial lines to the apex *H*.   With *H* as center and radii equal to $H\text{-}1^2$, $H\text{-}3^1$ and $H\text{-}2^2\text{-}4^2$, draw arcs intersecting similar radial lines in the pattern *P*, as shown.   Connect the various points, and the desired pattern is obtained.

Problem 59.   A Cone Intersected by a Cylinder Placed Vertically

Figure 96 illustrates the principles for developing a vertical cylinder intersecting with a cone.   The construction of this problem must be followed very carefully, as several of the operations are necessarily made over one another on the drawing, and the student must be careful to distinguish each process.

First draw the elevation of the cone *ABC*, in accordance with the dimensions given on the drawing.   The plan is then drawn, as shown, and thru the center *a* draw the diameter *FR*.   Establish the location of the center of the cylinder at *g*, and with *g* as center, describe a circle, which will represent the outline of the vertical pipe, of the diameter indicated on the drawing. Divide the upper half of the circle into equal spaces, as shown by the figures *1–4–7*.   Thru each of these points on the circle, from *a* as center, draw the half circles, as shown, intersecting the center line *FR* from *7* to *1*.

These half circles represent the plan views of horizontal planes which are projected to the elevation.   Then from points *7* to *1* on the center line *FR*, draw vertical lines intersecting the side of the cone *AC* from *1'* to *7'*.

From each of the points of intersection with the side of the

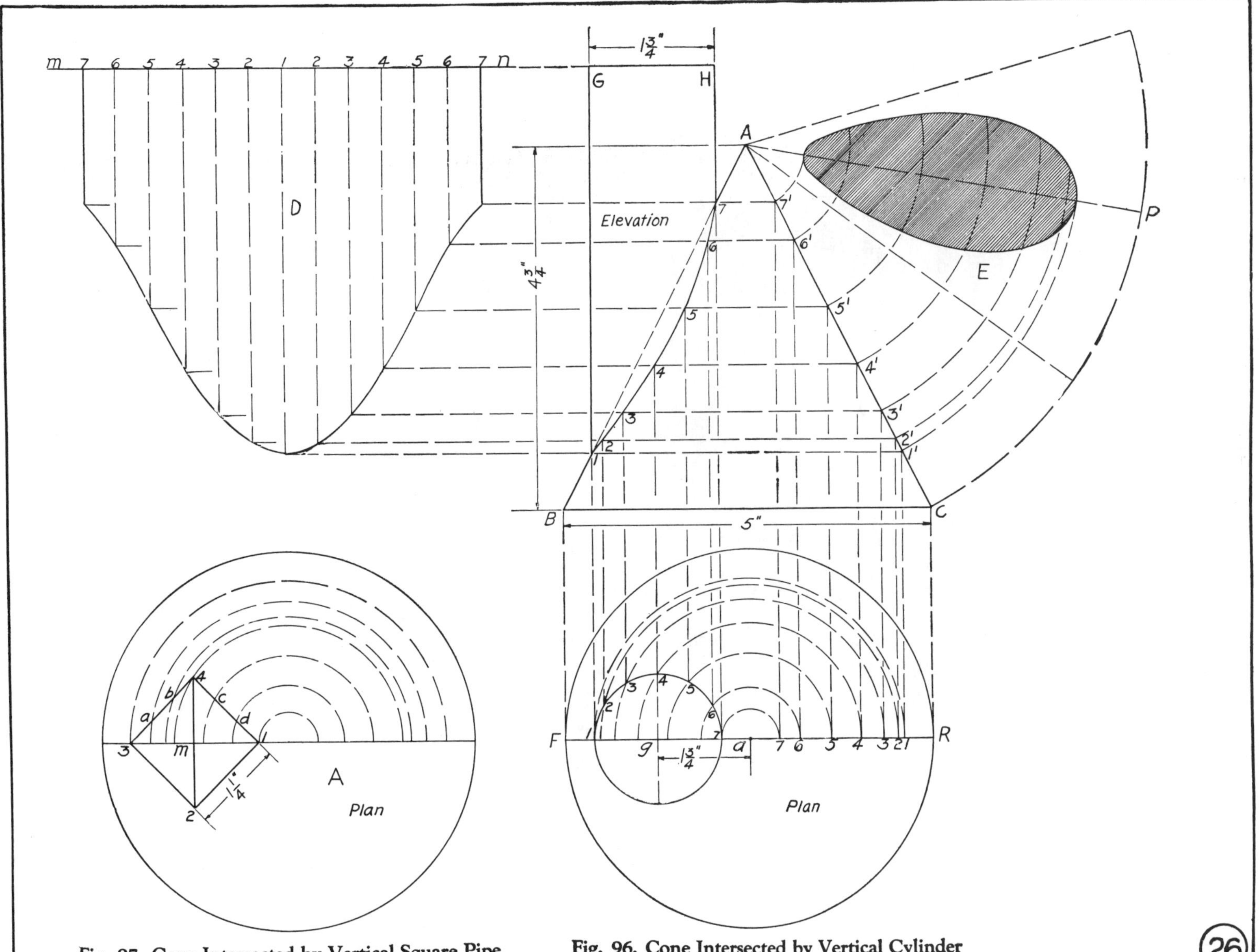

Fig. 97. Cone Intersected by Vertical Square Pipe

Fig. 96. Cone Intersected by Vertical Cylinder

cone, draw horizontal lines which are intersected by vertical lines drawn from similarly numbered points on the outline of the cylinder in plan.   A line traced thru these points, as shown, will be the miter line.

From the points *1* and *7* on the miter line, draw the vertical lines *1–G* and *7–H*, which connect from *G* to *H*, completing the elevation of the cylinder.

The pattern for the cylinder, shown at *D*, is obtained in the manner usual with all parallel forms; the stretchout is taken from the spaces upon the outline of the cylinder in plan, which are transferred to the line *mn*, as shown.   The method for obtaining the pattern for the cone is similar in process to Figure 80.

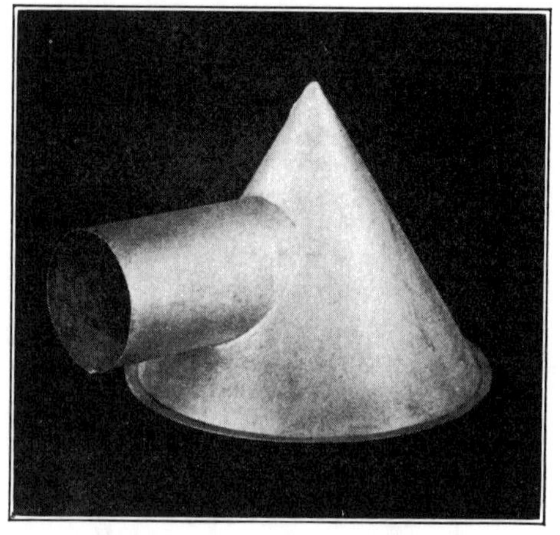

Problem 61.   Cone Intersected by Cylinder at Right Angles to its Axis.

A section of the pattern, and the method for obtaining the opening to fit the cylinder, is shown at *E*.

With *A* as center and radii equal to *A–7'*, *A–6'*, *A–5'*, *A–4'*, *A–3'*, *A–2'* and *A–1'*, describe arcs, as shown.   Draw the center line *AP*.   Then, measuring from the center line *FR* in plan, take the various distances along the arcs from the center line to the points on the outline of the small circle, representing the cylinder in plan, and place them on similar arcs in the pattern *E*, measuring on both sides of the line *AP*.   A line traced thru these points, as shown, will give the pattern for the opening in the cone.

Problem 60.   A Cone Intersected by a Vertical, Square Pipe

Using the plan shown at *A*, Figure 97, and the same size elevation as in Figure 96, develop the pattern for a cone intersected by a vertical square pipe placed diagonally, as shown in the plan at *m*.

The principles used for developing the patterns in Figure 96 are also applicable to various problems, no matter what the profile or form of the pipe may be.

Problem 61.   A Cone Intersected by a Cylinder at Right Angles to Its Axis

Figure 98.   The principles in this problem do not differ from those given in Figure 96.   Draw the elevation of the cone *ABC* and the plan view *F*.   Next, draw the elevation of the cylinder *G–3–3–H*; also its section, shown by *R*, which is divided into equal spaces, shown from *1* to *5*.   From these points, draw horizontal lines intersecting the side of the cone *AC*.

Then, from the points of intersection on the line *AC*, draw vertical lines intersecting the center line *ab* in plan at *2–4*, *1–5* and *2–4*.   With *F* as center and radii equal to *F–2–4*, *F–1–5* and *F–2–4*, describe the circles shown.   Extend the center line *ab* in plan and draw a duplicate of profile *R*, as shown by *D*, changing the position of the numbers shown from *1* to *5* to *1*.   From the points on profile *D*, draw horizontal lines intersecting similarly numbered circles, as shown by *1'* to *5'* in plan.   From these intersections, draw vertical lines, which intersect similarly numbered horizontal lines in the elevation.   A line traced thru these points will give the miter line between the cone and cylinder.

The pattern for the cylinder can be developed from the intersections in the plan or elevation in the usual manner.   The pattern for the opening in the cone is shown at *K*, and is obtained

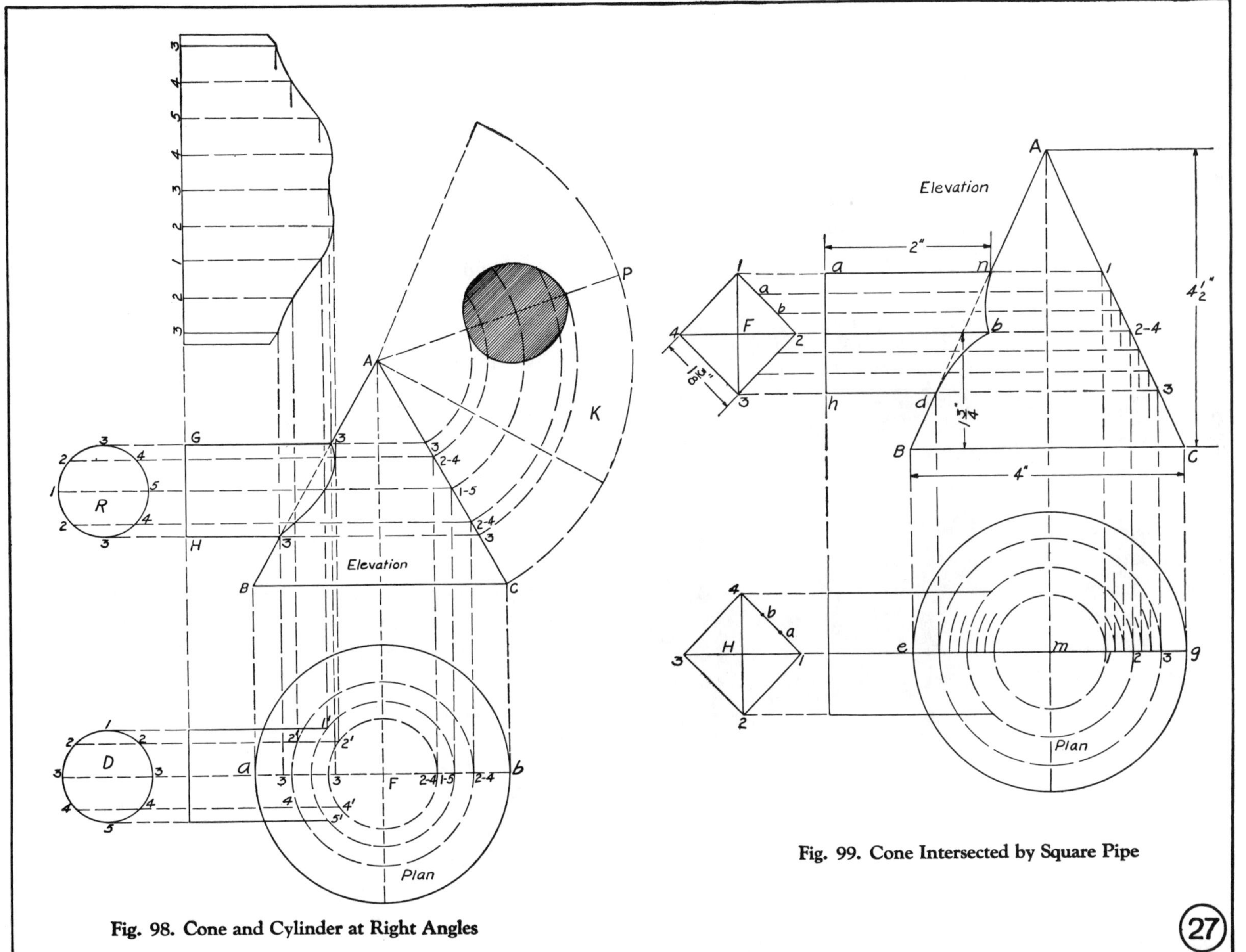

Fig. 98. Cone and Cylinder at Right Angles

Fig. 99. Cone Intersected by Square Pipe

27

in the manner explained in Problem 59. Measuring on either side of the center line *ab*, take the distances along the arcs in plan and place them on similar arcs in the pattern, measuring from the center line *AP*. A line traced thru the points thus obtained will give the pattern for the required opening in the cone.

Problem 62. A Cone Intersected by a Square Pipe Placed in a Horizontal Position

Figure 99 shows the plan and elevation of a cone intersected by a square pipe at right angles to its axis.

The principles illustrated in Figure 98 for obtaining the intersections between a cone and cylinder placed horizontally are also applicable to problems where the pipe is square or elliptical in form. Draw the plan view of the cone, and thru the center *m* draw the diameter *eg*. Directly above the plan, draw the elevation *ABC*, and in its proper position, the elevation of the square pipe *ahdn*; also the profile shown by *F*, which divide into equal spaces to obtain the points *a* and *b*.

These points are also located upon the profile *H* in plan, and are projected to the plan and elevation of the cone in the

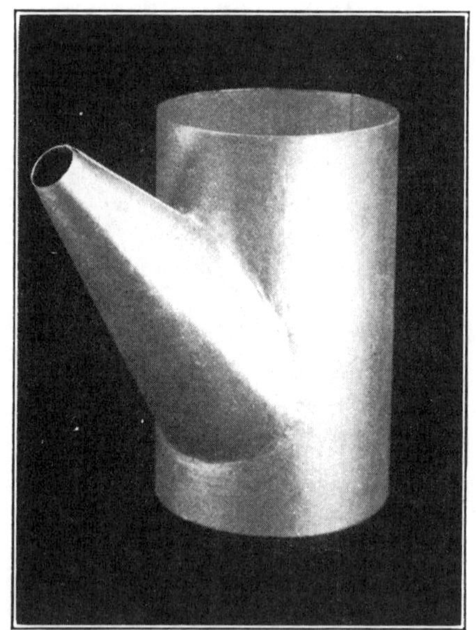

Problem 63.   Frustum of Cone Intersecting a Cylinder Obliquely.

same manner as the points on the profile of the cylinder in Figure 98.

Obtain the miter line *nbd* in the elevation formed by the junction of the square pipe and the cone, and develop the patterns for the cone and the horizontal square pipe.

Problem 63. The Frustum of a Cone Intersecting a Cylinder Obliquely

The principles given in this problem are applicable to cone and pipe intersections at any angle, and the cone may be placed in the center or to one side of the pipe. This method can also be applied to problems where the vertical pipe is square or rectangular in form.

Figure 100 shows the plan, elevation and patterns for the frustum of a cone intersecting a cylinder of greater diameter than itself at an angle of 45°. The drawing is made to a scale of 6 inches to 1 foot.

Draw the elevation of the cylinder *AFCB*, and the plan view, shown at *G*. Locate the point *R* and draw the elevation of the cone *Rba*, extending the sides of the cone until the base line *7–1* is obtained. From the center *e* on the base line, draw the half view of the base, which divide into equal spaces, as shown from *1* to *7*. From the various points on the half view of the base and parallel to the center line of the cone, erect lines intersecting the base of the cone, and from these intersections draw radial lines to the apex *R*.

Thru the center of the cylinder in plan, parallel to the base line *BC* in elevation, draw the line *R'H*, which is intersected at *R'* by a vertical line projected from *R* in the elevation.

With *m* as center, draw a full view of the base of the cone, and number the divisions from *1* to *7*, as shown in *D*.

From these points, draw horizontal lines, which intersect by

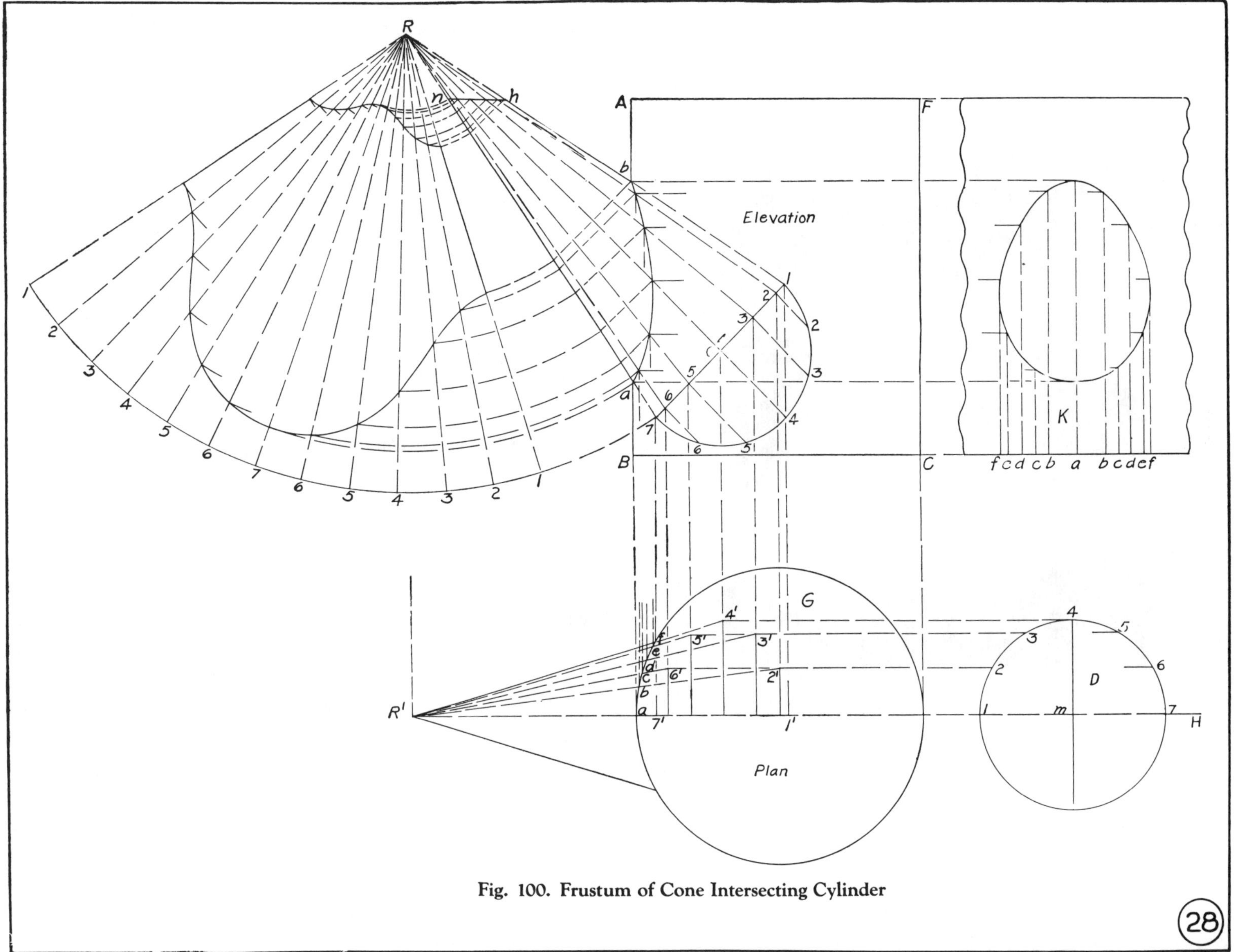

Fig. 100. Frustum of Cone Intersecting Cylinder

vertical lines drawn from similarly numbered intersections on the base line *7–e–1* in the elevation.

From these intersections in plan, shown by *1'–2'–3'*, etc., draw lines to the apex *R'*, intersecting the plan of the cylinder

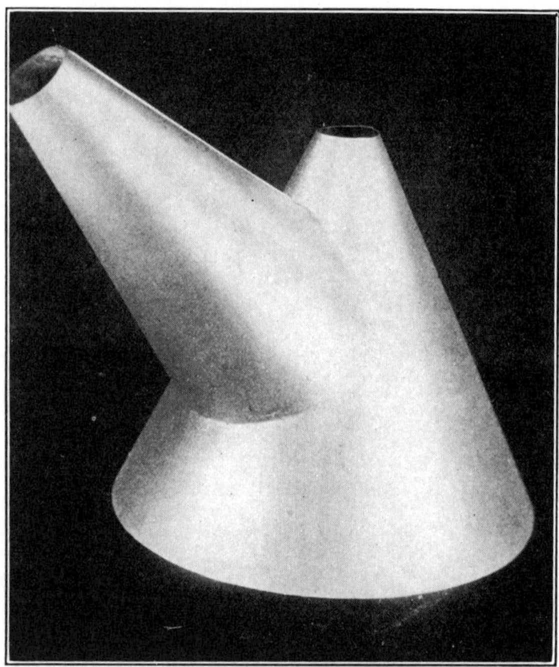

Problem 64.   Frustum of Cone Intersecting
a Cone Obliquely.

from *a* to *f*. From the intersections *a* to *f*, draw vertical lines which intersect corresponding radial lines in the elevation. A line traced thru these points, as shown, gives the line of intersection or miter line between the cone and cylinder.

The points of intersection on the miter line are now projected at right angles to the center line of the cone to the true edge line *Ra,* as are also the points at the intersection of the radial lines

with the horizontal line *nh* which represents the upper base of the frustum.   With *R* as center and *R–7* as radius, describe the stretchout arc, and develop the pattern for the frustum of the cone in the regular way.

The pattern for the opening in the cylinder, shown at *K*, is obtained by the parallel-line method; the stretchout line *BC* extended is drawn in the position shown, and the spaces *a, b, c, d, e* and *f*, as found in the plan, are spaced off on that line.   Thru the points thus located, vertical lines are now drawn, and intersected by horizontal lines from similarly numbered points in the line of intersection in the elevation.

Problem 64. The Frustum of a Cone Intersecting a Cone of Unequal Diameter Obliquely

Figure 101 shows the method of obtaining the patterns for intersecting cones, and the drawing is made to a scale of 4 inches to 1 foot.   Draw the elevation of the larger cone as *ABC*, and in its proper position below same, describe the circle *E*, which represents the plan view of the base.   Next, thru point *b* on the line *AB*, draw the center line of the smaller cone at an angle of 45° with the base of the larger cone, and locate the points *F* and *m*.

Thru *m* at right angles to the center line *Fm*, draw the line *5–m–1*.   With *m* as center, describe the semi-circle *1–3–5*, which represents a half view of the base of the smaller cone. Divide the outline into a convenient number of equal parts, shown from *1* to *5*, and from these points draw lines to the base line *5–m–1*.   Radial lines are now drawn from these intersections to the apex *F*.   Divide the quarter-circle *6–8* in plan *E* into equal spaces, as shown by the figures *6–7–8*, and draw radial lines to the center *n*, as shown.   In practical work, it will be found necessary to use a larger number of spaces in order that the line of intersection may be more accurately traced.   From

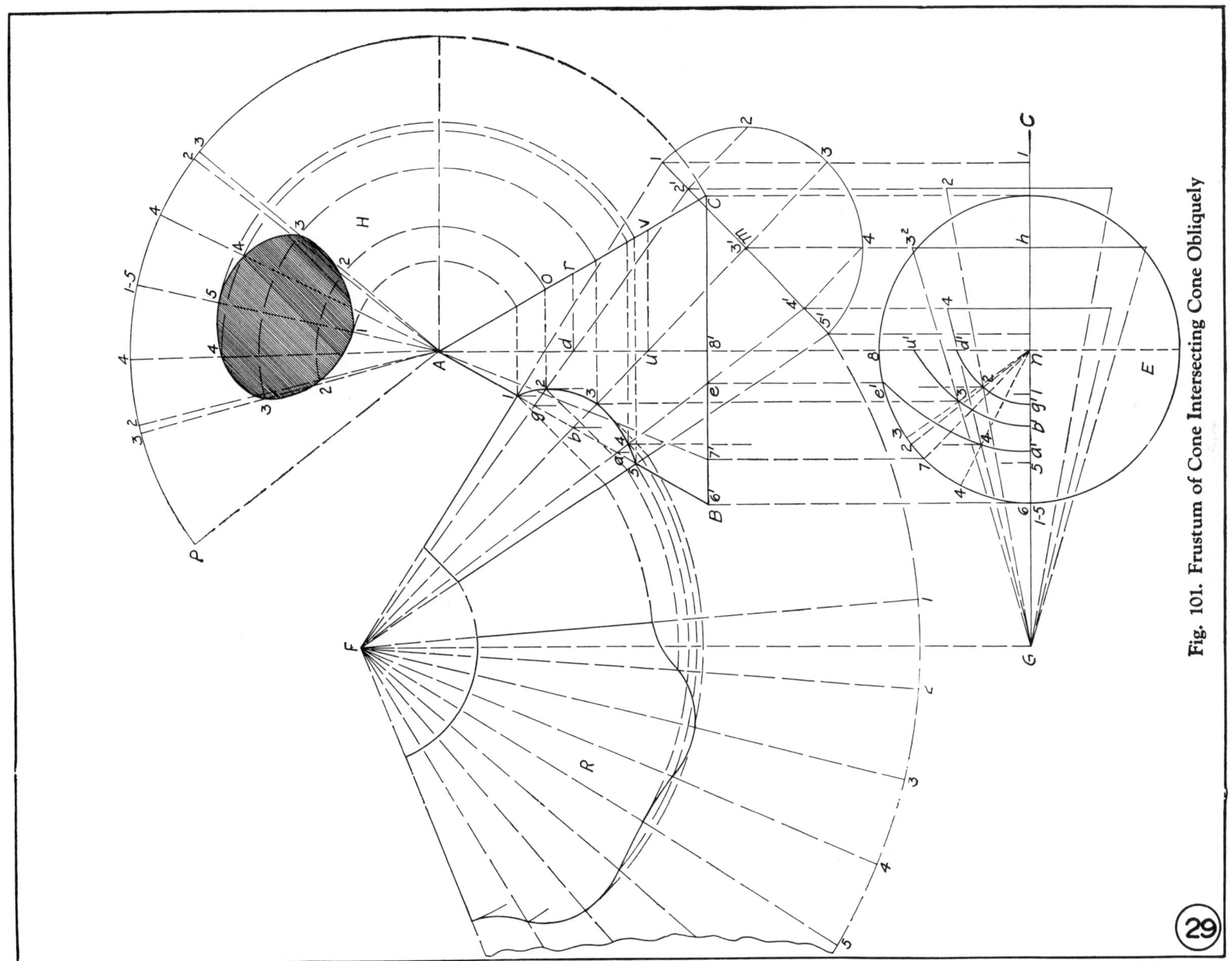

Fig. 101. Frustum of Cone Intersecting Cone Obliquely

the intersections *6–7–8* in plan, draw vertical lines intersecting the base line *BC*, from which radial lines are drawn to the apex *A*, intersecting the radial lines of the smaller cone *F–5–1*, as shown.

A series of sections of both cones are next drawn in the plan. The sections of the smaller cone will, in each case, be a triangle, while the sectional curves of the larger cone will be elliptical, parabolic or hyperbolic curves, as the case may be. From the point of intersection with the side of each triangle and its corresponding sectional curve, vertical lines are drawn which intersect similarly numbered radial lines of the smaller cone, and the line of intersection between the two cones is traced thru these points.

The sectional triangles of the smaller cone are obtained in the following manner: From the points *5, 4', 3', 2'* and *1*, on the base line of the smaller cone, draw vertical lines to the plan, and set off distances from the center line *GC* similar to the distances from *5–m–1* in the half view of the base; that is, make *3'–3* equal $h–3^2$, etc. From these points, draw lines to the apex *G*, completing the triangles.

The next step is to obtain the sectional curves of the larger cone on the planes *F–4'*, *F–3'* and *F–2'*. The extreme upper and lower points *1* and *5* are the points of intersection of the central section, and are projected directly to the plan from the elevation. Where the plane *F–4'* crosses the base line *BC* at *e* and the radial lines drawn from *6'* and *7'*, shown by *abg*, lines are projected into the plan until they intersect similar radial lines *6* and *7*; also the plan of the base at *e'*. The irregular curve *a'e'*

traced thru these points of intersection is a section of the larger cone produced by the intersection of the cutting plane *ea* in elevation.

In similar manner obtain the sectional curves thru the planes *bu* and *gd*, as shown in plan by *b'u'* and *g'd'*, the distance from *n* to *d'*, and *n* to *u'* being obtained from *u* to *v* and *d* to *r*, respectively, in the elevation.

From the points of intersection in plan where the sectional triangles of the smaller cone intersect the sectional curves of the larger cone, vertical lines are drawn which intersect similar radial lines in the elevation at *1, 2, 3,4* and *5*, as shown.

From these intersections at right angles to the center line *Fm*, draw lines intersecting the side of the cone *F–5*.

With *F* as center and *F–5'* as radius, describe the stretchout arc and develop the pattern for the frustum of the cone in the usual manner, shown at *R*.

A section of the pattern and the method for obtaining the opening in the larger cone is shown at *H*.

With *A* as center and *AC* as radius, describe the arc *CP*. Next, draw any line as *A–1–5*, on either side of which place the various divisions *4, 2, 3*, which are taken from the arc *6–8* in plan and placed upon the arc *CP*, as shown.

From these points, draw radial lines to the apex *A*, which intersect by arcs drawn from *A* as center with radii equal to the divisions on the side of the cone, as shown on the line *AC*.

A line traced thru the intersections shown from *1* to *5*, will give the pattern for the opening in the larger cone.

CHAPTER VII
TRIANGULATION

There are numerous irregular forms arising in sheet-metal work, the patterns for which cannot be developed by the regular methods employed in the two previous chapters. These irregular shapes are so formed that, altho straight lines can be drawn upon them, the lines would not run parallel to one another, nor would they all incline to a common center. In parallel-line developments, the distance between any two lines running with the form is the same at both ends of the article, while in radial developments, all lines running with the form tend toward a common center or apex, so that the distances between such lines at one end of the article (provided it does not reach to the apex) govern those at the other end.

Hence, in the development of the pattern for any irregular form, it becomes necessary to drop all previous methods, and simply proceed to divide the drawing representing the surface of the article into triangles.

Then from the drawing, the true lengths of the various sides must be found, and the triangles constructed therefrom.

The construction of a triangle whose three sides are given is not a difficult problem, and it becomes a simple problem in geometry to construct the triangle. Having found the true lengths of the sides of such triangles, reproduce them in regular order in the pattern, and, hence, the term triangulation is most fittingly applied to this method of development of surfaces.

Problem 65.   Transition Piece, Square to Square

While irregular forms are largely curved surfaces, the method of triangulation is best illustrated by its application to a form having plane surfaces, as shown in Figure 102. Both bases are square, and, in this case, parallel, but diagonally-arranged in their relation to each other, as may be seen from the drawing. Draw the elevation and plan view, in which a, b, c, d represents the square base, and 1, 2, 3, 4 the plan view of the square top, each side of which shows its true length. Now, connect the top and base by drawing lines from the corners in plan, as shown. These lines represent the bases of the triangles, which must be constructed so as to find the true lengths of these lines.   This is accomplished by constructing, in each case, a right-angled triangle, whose base is equal to the length of any fore-shortened line in the plan, and its altitude to the vertical height of the same line shown in the elevation.   The hypotenuse of such a triangle will then be equal to the true length of the line.   In this case, the lines b–1, C–2, d–3, a–4, etc., are all represented by lines of the same length. The vertical height AB is the same in the case of each line.

A triangle constructed by the above method will be sufficient to indicate the true length of these lines, and is constructed as follows: Draw any horizontal line as mn, and from m erect the perpendicular mh equal to the altitude AB in the elevation.

Problem 65.   Transition Piece, Square to Square.

As all of the lines connecting the corners in plan are equal, only one triangle is necessary. Therefore, take the distance from b to 1 in plan and place it, as shown, from m to n, and draw the line hn, which represents the true length of the lines b–1, c–2, d–3, etc., in plan.

The pattern is to be laid out in one piece, with a seam thru 1–e in plan.

Take this distance 1–e and place it, as shown, from m to g, and draw a line from h to g, which will be the true length of the seam line 1–e.   The true lengths of all the lines in plan having been found, the triangles may now be placed in position in the pattern, care being observed that the adjacent triangles are completed in the same order as they are shown on the solid.

Draw any horizontal line as cd in the pattern, equal to cd in the plan.   Now, with radius equal to hn in the triangles and cd in the pattern as centers, describe arcs intersecting each other at 3.   Draw lines from c to 3 and 3 to d.   Then c–d–3 is the correct development of the surface c–3–d in plan.   The adjacent triangles c–2–3 and d–3–4 may next be constructed.   With 3–2 in the plan as radius, and 3 in the pattern as center, describe the arcs 2 and 4; these arcs are then intersected by arcs described from c and d as centers, with a radius equal to hn in the diagram of triangles, thus developing the triangles c–3–2 and d–3–4, which correspond to similarly numbered surfaces in the plan. Now, with radius equal to cb and da in plan, and c and d in pattern as centers, describe the arcs b and a, which intersect by arcs described from 2 and 4 as centers and hn in the triangles as radius.   Draw lines from 2 to b to c and 4 to a to d in the pattern, which is the pattern for the sides d–4–a and c–2–b in plan.   In like manner develop the surface of the figure shown by 4–a–1 and 2–b–1 in the plan.   Then, with radius equal to ae in plan, and a and b in the pattern as centers, describe the arcs mm, which intersect by arcs described from 1 and 1 as centers and hg in the diagram of triangles as radius.   Draw the lines b to m to 1 and a to m to 1, completing the pattern.

Problem 66.   Register Box, Rectangular to Round

Figure 103 shows the manner of developing the pattern for a register box whose top is rectangular and base is round, and may also be described as a transition piece; that is, a form used to connect outlines unlike in shape.   A fitting of this kind is frequently used in the sheet-metal trades, particularly in the construction of bases for chimney tops, ventilator heads and fan connections in blow-pipe work.

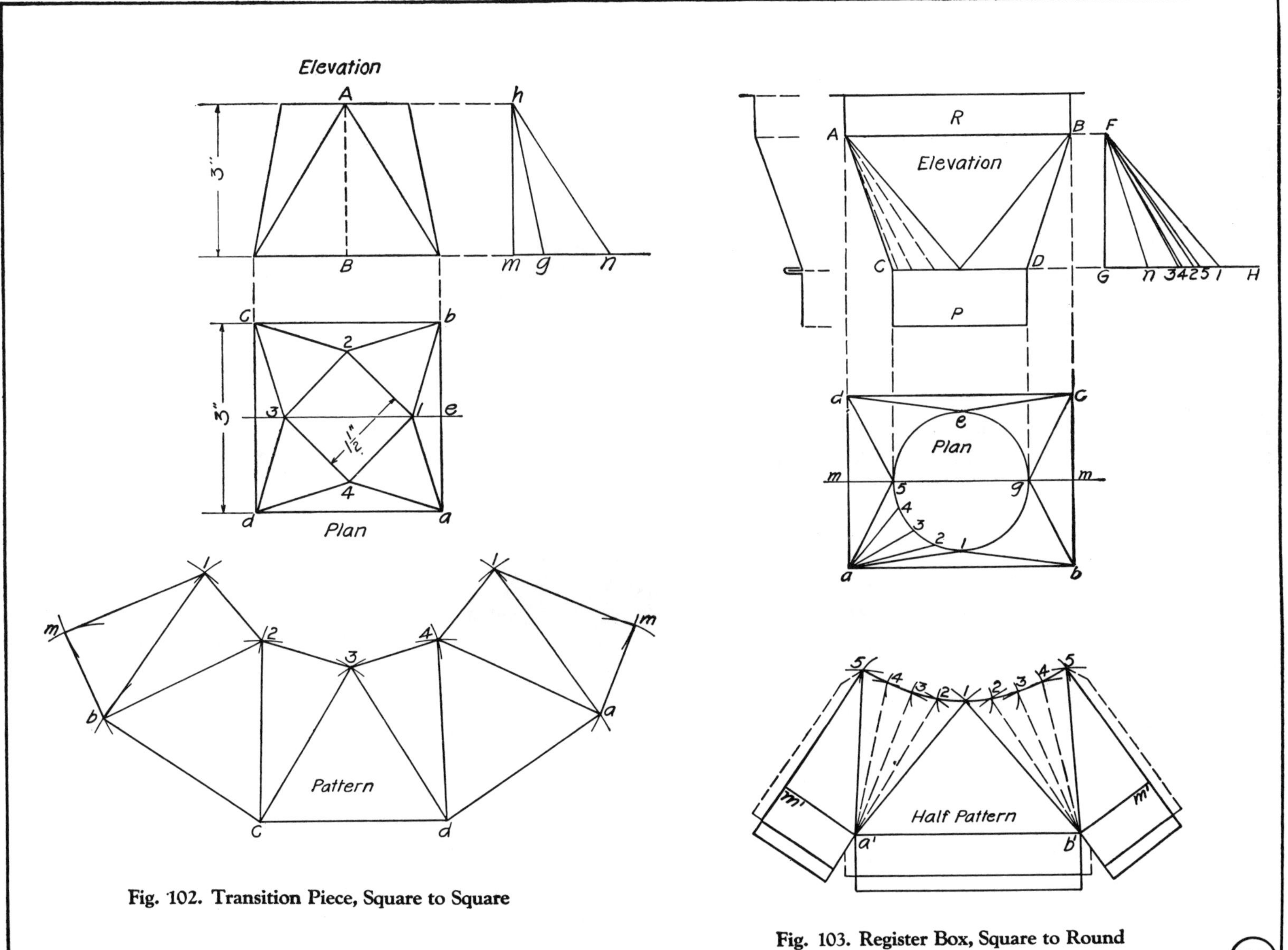

Fig. 102. Transition Piece, Square to Square

Fig. 103. Register Box, Square to Round

30

The elevation, plan and half pattern shown in Figure 103 are drawn to a scale of 3 inches to the foot. First, draw the rectangle *abcd*, which represents the plan of the top, and circle *e–5–1–g*, which shows the plan of the base of the article. As the circle is in the center of the rectangle, making the four quarters symmetrical, it is necessary only to divide the one-quarter circle into a number of equal parts, as shown by the figures *1, 2, 3, 4, 5,* from which draw lines to the corner *a*. These lines will form the bases of a series of triangles whose altitude is equal to the vertical height of the article, and whose hypotenuses will be the real distances from *a* in the base to the points assumed in the curve in the top.

Next, draw the elevation *ABCD*, and add the straight flange and collar to the top and base, as shown. The next step is to find the true lengths of the lines *1–a, 2–a, 3–a*, etc., in plan. To construct a diagram of triangles, first draw the line *G H*, and from *G* lay off the distances, shown by the lines in plan, thus making *G–1* equal to *a–1*, *G–2* equal to *a–2*, etc. At right angles to *HG*, draw *GF*, in height equal to the straight height of the article, as shown in the elevation, and connect the points on the line *G H* and *F*. Also set off the distance *m–5* from *G*, and draw the line *F–n*, which will give the true length of the seam line *m–5* in plan.

To develop the half pattern, first draw any line, as *a'b'*, equal in length to *ab* in plan. Now, with *F–1* of the diagram of triangles as radius and *a'* and *b'* as centers, describe arcs intersecting each other, thus establishing point *1* of pattern. Next, with the points *a'* and *b'* as common centers and radii equal

to the true lengths of the lines *a–1, a–2, a–3*, etc., of the plan, as shown in the diagram of triangles, describe arcs of indefinite length. Set the dividers to the length of one of the spaces on the quarter circle in plan, and, commencing at the point *1* in pattern, step off a number of spaces on each side to correspond to those shown on the quarter-circle from *1* to *5* in plan. Thru the points thus obtained, trace a line, as shown from *5* to *6*. With *a'* and *b'* of pattern as centers, and *am* of plan as radius, describe short arcs, which intersect other arcs, described from *5* and *5* as centers, and *Fn* of diagram of triangles as radius, thus establishing the points *m'* of the pattern.

Now, connect the various points by drawing lines from *5* to *m'* to *a'* and *5* to *m'* to *b'*, completing the half pattern for the tapering body of the register box, shown by *ABCD* in the elevation. The vertical flange *R* and laps for seaming are added, as shown in the pattern. The lower collar, shown at *P*, is simply a straight piece of pipe for which a pattern is not required, as it can be laid out directly upon the metal.

Problem 66. Register Box, Rectangular to Round.

Problem 67. Register Box, Rectangular to Round, Vertical on Two Sides

Figure 104 illustrates a condition occasionally met by the sheet-metal worker engaged in erecting stacks, furnace hoods, register boxes, and various other fittings used in heating and ventilating work. The principles in this problem do not differ from those given in the preceding problem, and are applicable,

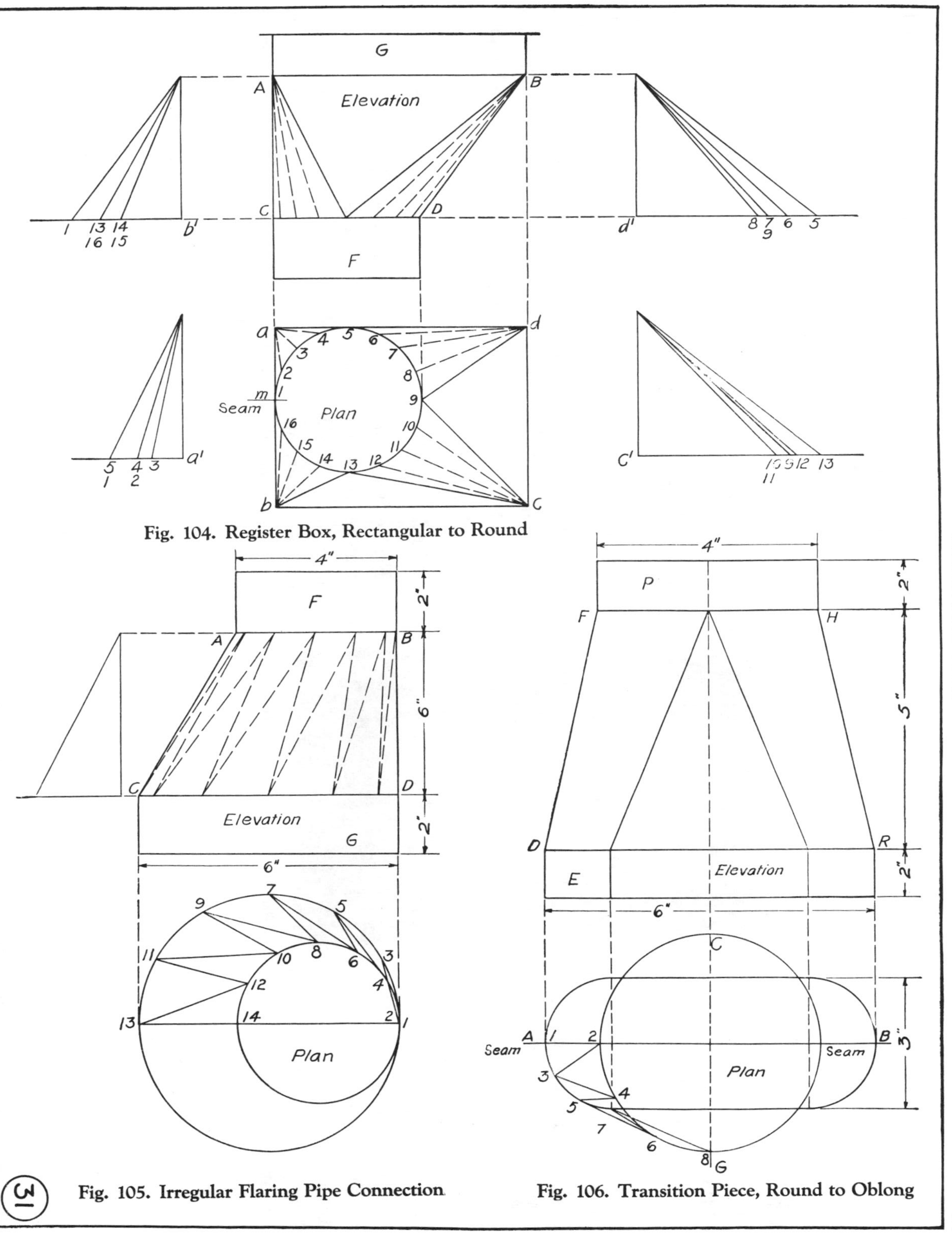
Elevation
Seam
Plan
Fig. 104. Register Box, Rectangular to Round
Elevation
Plan
Fig. 105. Irregular Flaring Pipe Connection
Elevation
Seam
Plan
Seam
Fig. 106. Transition Piece, Round to Oblong
31

whether the round opening of the article is placed exactly in the center of the base or at one side or corner.

Figure 104 shows a drawing (3 inches to the foot) of a register box whose top is rectangular, and whose base is round and placed in one corner of the rectangle, as shown. First, draw the rectangle *abcd*, which represents a plan view of the top, and the circle *1–5–9–13*, which shows the plan of the round base of the article. As the circle is not in the center of the rectangle as in the previous problem, the four quarters are not symmetrical, and it is necessary to divide the entire circle into a number of equal parts, as shown, from which draw lines to the corners *a, b, c* and *d*. Next, draw the elevation *ABCD* and add the straight flange *G* and the round collar *F*, as shown. The true lengths of the lines in each quarter of the plan are found by constructing the diagram of triangles, shown at *a′, b′, c′* and *d′*, by the method described in the previous problem. Having found the true lengths of all lines in plan, develop the full pattern for register box, placing the seam at *m–1*, as shown in plan.

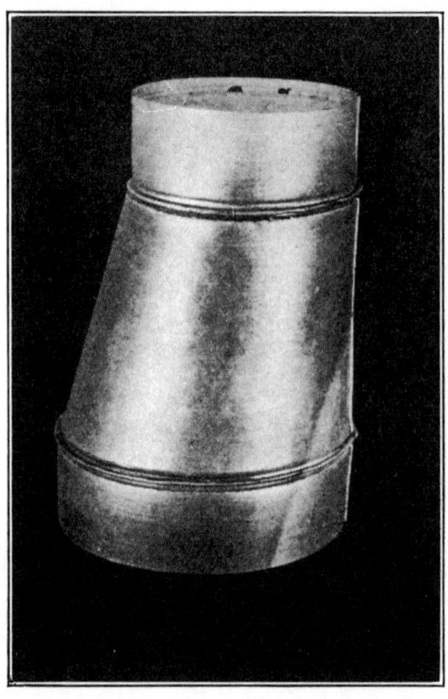

Problem 68. Irregular Flaring Pipe Connection.

Problem 68. Irregular Flaring Pipe Connection, Vertical on One Side

The style of pipe-fitting shown in Figure 105 is frequently used in blow-pipe-fitter's work, especially where the lower side of a round duct is required to be perfectly straight, to avoid any obstruction to the material that is being forced thru the pipe by means of a fan.

Draw the plan and elevation in accordance with the dimensions given. In the plan view, after drawing the horizontal center line, shown by *13–1*, divide the outline of each of the half sections into the same number of equal parts. Number these points, as shown, from *1* to *13*, and draw lines connecting the successive points. The true lengths of all lines in the plan must now be determined by means of a diagram of triangles, and the development of the pattern for the irregular flaring section, shown by *ABCD* in the elevation, will differ in no material respect from the development explained in Problem 66. The sections *F* and *G* are simply straight pieces of pipe for which a pattern is not required.

Problem 69. Transition Piece, Round to Oblong

Figure 106 shows the plan view and side elevation of a pipe fitting or transition piece used in connecting an oblong and round pipe. The sheet-metal worker engaged in heating and ventilating work has frequent use for a fitting of this kind, and it is generally known as a center boot, which is used to connect an oblong riser or wall pipe with a round horizontal pipe. The round end of the boot is connected to the pipe by means of a three- or four-pieced elbow which gives a gradual turn, making direct connection with the furnace hood or air chamber in hot-air heating work.

In constructing this drawing, the plan is to be drawn first. Draw the center line *AB* and construct the outlines of the upper and lower bases. Draw the vertical center line *CG*, which divides the plan into symmetrical quarters; all the work that is

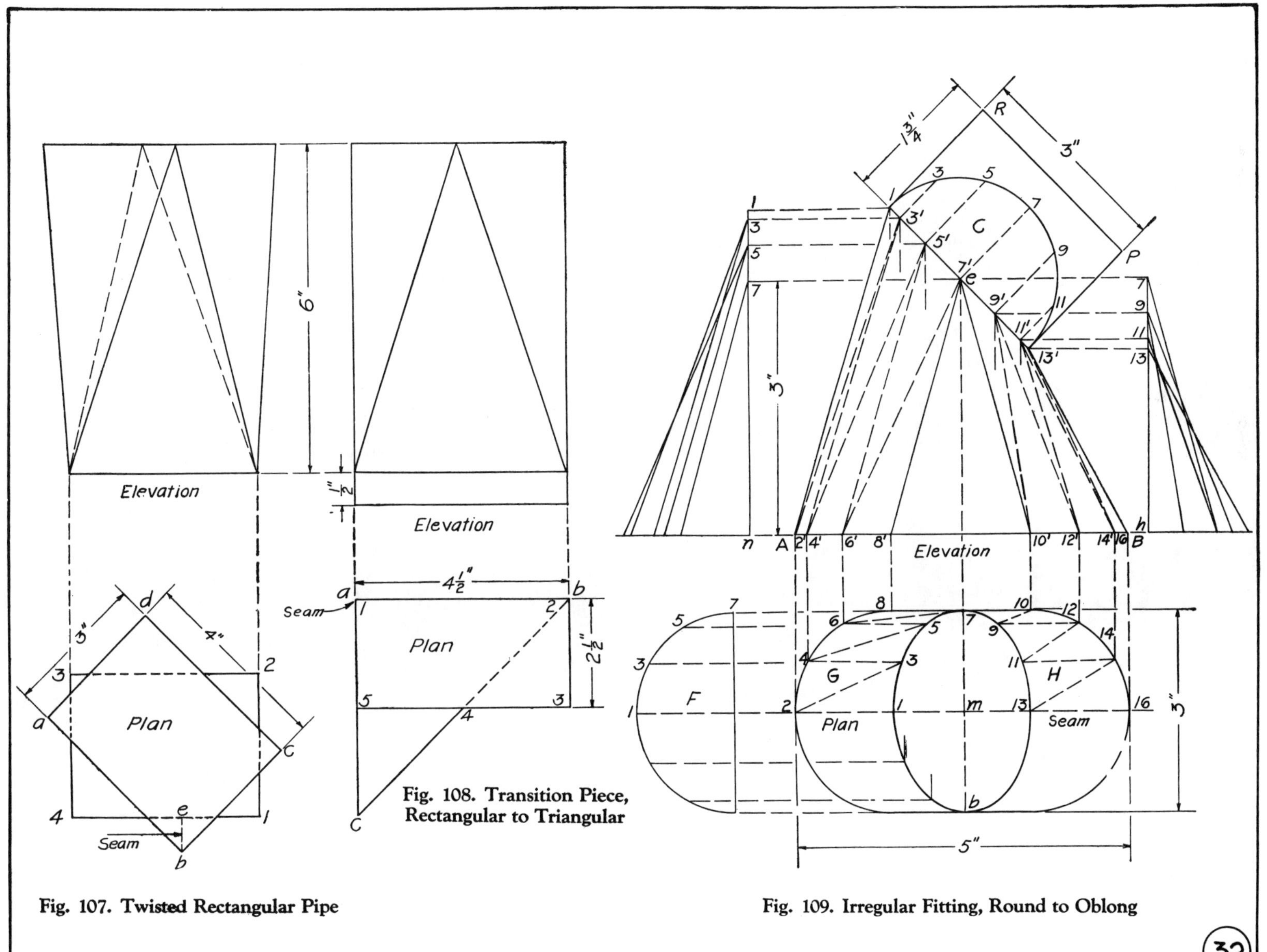

Fig. 107. Twisted Rectangular Pipe

Fig. 108. Transition Piece, Rectangular to Triangular

Fig. 109. Irregular Fitting, Round to Oblong

32

necessary for the development of the pattern may, therefore, be accomplished in one of these divisions. Next, draw the side elevation of the transition piece, as shown by $FHDR$, and place the straight collars E and P in position, as shown. Divide the quarter-circle representing the upper base into a number of equal parts, shown by $2, 4, 6, 8$, and an equal number of spaces are set off on the quarter-circle representing the lower base, shown by $1, 3, 5, 7$. Draw lines connecting the successive points in the plan, shown by $1-2$, $2-3, 3-4$, etc. The true lengths of these lines are next determined by means of a diagram of triangles constructed by the method previously shown for such cases, and a one-half pattern is developed, placing the seams in the position shown by A and B in the plan. The student being already familiar with the method of procedure, no further instruction is necessary.

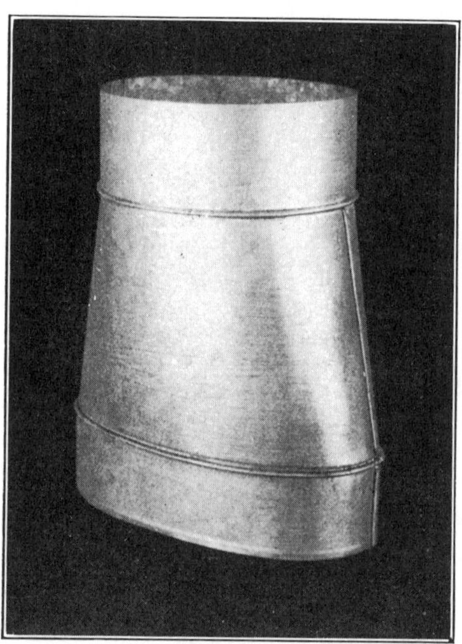

Problem 69.   Transition Piece, Round to Oblong.

Problem 70.   Twisted Rectangular Pipe

An elevation and plan view of a twisted rectangular pipe is shown in Figure 107. The profiles of the upper and lower ends are alike, and they are placed diagonally one above the other, as shown in the plan. In order that the plan be more readily understood, the plan of the bottom of the pipe is represented by the numbers $1, 2, 3$ and 4, and the top end by the letters a, b, c and d. Connect the various corners and find the true lengths of the lines by the usual method that has been explained in similar triangulation problems already given. Develop the pattern in one piece, placing the seam in the position shown by the line eb in the plan view.

Problem 70.   Twisted Rectangular Pipe.

Problem 71.   Transition Piece, Rectangular to Triangular

Figure 108 shows a transition piece, the bottom of which is triangular in form, the top being a rectangle. Let $1, 2, 3, 4$ and 5 represent the plan of the top, and a, b and c the plan of the bottom of the transition piece. Draw the plan and elevation and develop the pattern in one piece, placing the seam at the corner, shown by $c-1$ in the plan.

Problem 72.   Irregular Fitting, Forming a Transition from Round to Oblong.   Upper Base Inclined 45°

In Figure 109, $AB-1-13$ represent the side elevation of an irregular fitting, whose top opening, shown by the line $1-13$, does not run parallel with the lower base. The first step is to draw a plan view showing the outline of the oblong base with semi-circular ends. Draw the center line $2-m-16$, thus dividing the plan into symmetrical halves.   From m draw a vertical line, and locate the point e three inches above the base line AB in the elevation.   Thru the point e, draw the line 1–13, the angle of inclination being 45°.   With e as center, describe the half-circle representing the half profile of the upper base.   Next, divide the half-profile c into a number of equal parts, and from these divisions at right angles to the upper base, draw lines intersecting the line 1–13, as shown.

Divide the lower base of the fitting, as shown in the plan, into a number of equal parts.   Next, take a tracing of the half profile C and place it in the position shown at F, numbering the divisions, as shown, from 1 to 7. From the points on the half-profile F, draw horizontal lines, which intersect by vertical lines drawn from the points 1–$3'$,$-5'$, etc., on the upper base in the elevation.   A line traced thru these points of intersection will give the foreshortened view of the upper base in plan. Draw lines connecting the points on the upper and lower bases in plan, as in the preceding problems, for the purpose of defining the triangles.   The true lengths of the lines 1–2, 2–3, 3–4, etc., are found in a slightly different manner from that used in the pre-

Problem 71.   Transition Piece, Rectangular to Triangular.

vious problems, in which the upper and lower bases were parallel with each other.   In this case, the vertical distances are not the same and the true lengths of the lines in plan are found in the following manner: Extend the lower base line AB in the elevation to the right and left, as shown on the drawing, and on these lines set off the lengths of the lines 1–2, 2–3, 3–4, etc., as they appear in the plan at G and H.   Draw the vertical lines 1–n and 7–h, and upon these lines the vertical heights are projected from the elevation, as shown.

The true lengths of all the lines having been determined, develop the full pattern, placing the seam on the line 13–16, as shown in the plan view. The collar, shown by 1–R–P–13, is simply a short piece of round pipe, the pattern being laid out directly upon the metal.

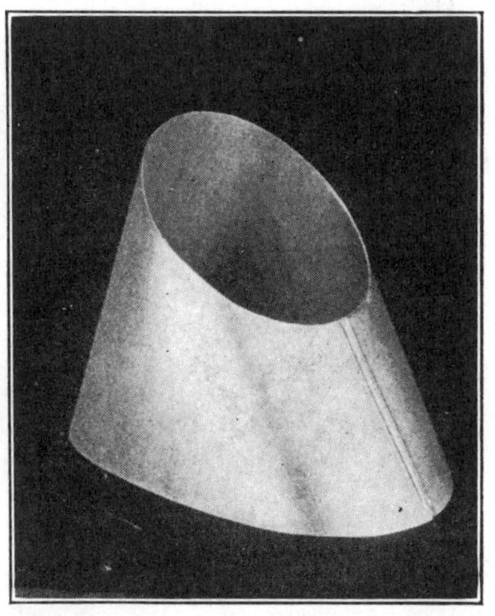

Problem 72. Irregular Fitting, Transition from Round to Oblong.

Problem 73.   Irregular T-Joint

In many cases of blow-piping, it is necessary to connect two pipes of different diameters at various angles.   An ordinary T-joint would not suffice, for the reason that it is important to secure an easy flow of air thru the pipes.   To accomplish this result, an irregular flaring connection piece is often used, as shown in Figure 110.   The upper base is round and is rightly inclined at an angle of 30°.   The lower base is oblong in form and is a portion of a cylinder.   It is desired to so construct the fitting that the opening in the large pipe will be somewhat larger than the diameter of the smaller pipe. As may be seen from Figure 110, the method of development does not differ materially from that of the preceding problem. First, draw the plan view of the oblong base, noting that, in this case, the outline is a foreshortened view of the real surface. Next, draw a line vertically upwards from *C*, the center of the plan, to the point *a* in the elevation.

Thru *a* draw the line *1–13* at an angle of 30° with the horizontal; with *a* as center and *a–1* as radius, describe the half circle representing the one-half full view of the upper base, as shown. With *b* on the vertical center line as center and a radius of 3-1/2 inches, describe in the elevation the arc *geh*, representing the view of the lower base and a section of the large pipe. Since the view of the oblong base in the plan is foreshortened, it is necessary to draw a full view, in order to ascertain the true distance around the lower base. To produce the full view, shown at *A*, extend the center line *2–16* in the plan toward the right, and on this line lay off the stretchout of the lower base, as shown by the numbers *2'*, *4'*, *6'*, *8'*, etc., on the arc *geh* in the elevation. From these points on the stretchout line in the full view, draw

Problem 73. Irregular T-Joint.

vertical lines, which intersect by horizontal lines drawn from similarly numbered points in the plan. A line traced thru these points will give the true outline of the lower base of the fitting; measurements are taken from points on this outline for the stretchout of the lower edge of the pattern, their true lengths thus being shown. The true lengths of these lines are found by constructing a diagram of triangles on both sides of the elevation. Two diagrams are constructed to avoid confusion from having a number of lines cross on the drawing. The true lengths of all lines having been found, the pattern may be developed by methods precisely like those used in preceding problems.

Problem 74. Roof Collar, Square to Round, Base Obliquely Inclined

Figure 111 illustrates this problem. In this figure, *1–9–a–b* is the elevation of a roof collar having a round top and square base, when viewed on a horizontal line. The collar fits over an inclined roof having a pitch of 45°, shown by the line *CG* in the elevation. First, draw the plan in accordance with the dimensions given on the drawing. Next, draw the elevation and construct the diagram of triangles, as shown at *e*. The true lengths of all the required distances may now be taken from their respective places on the drawing and a full pattern constructed in exactly the same manner as has been explained in similar problems already given.

Problem 75. Scalene or Oblique Cone

The development of this problem (shown in Figure 112) illustrates a short method of triangulation applicable to a number of problems, particularly to those represented by transition pieces, the bases of which can be inscribed within a circle. Figure 112

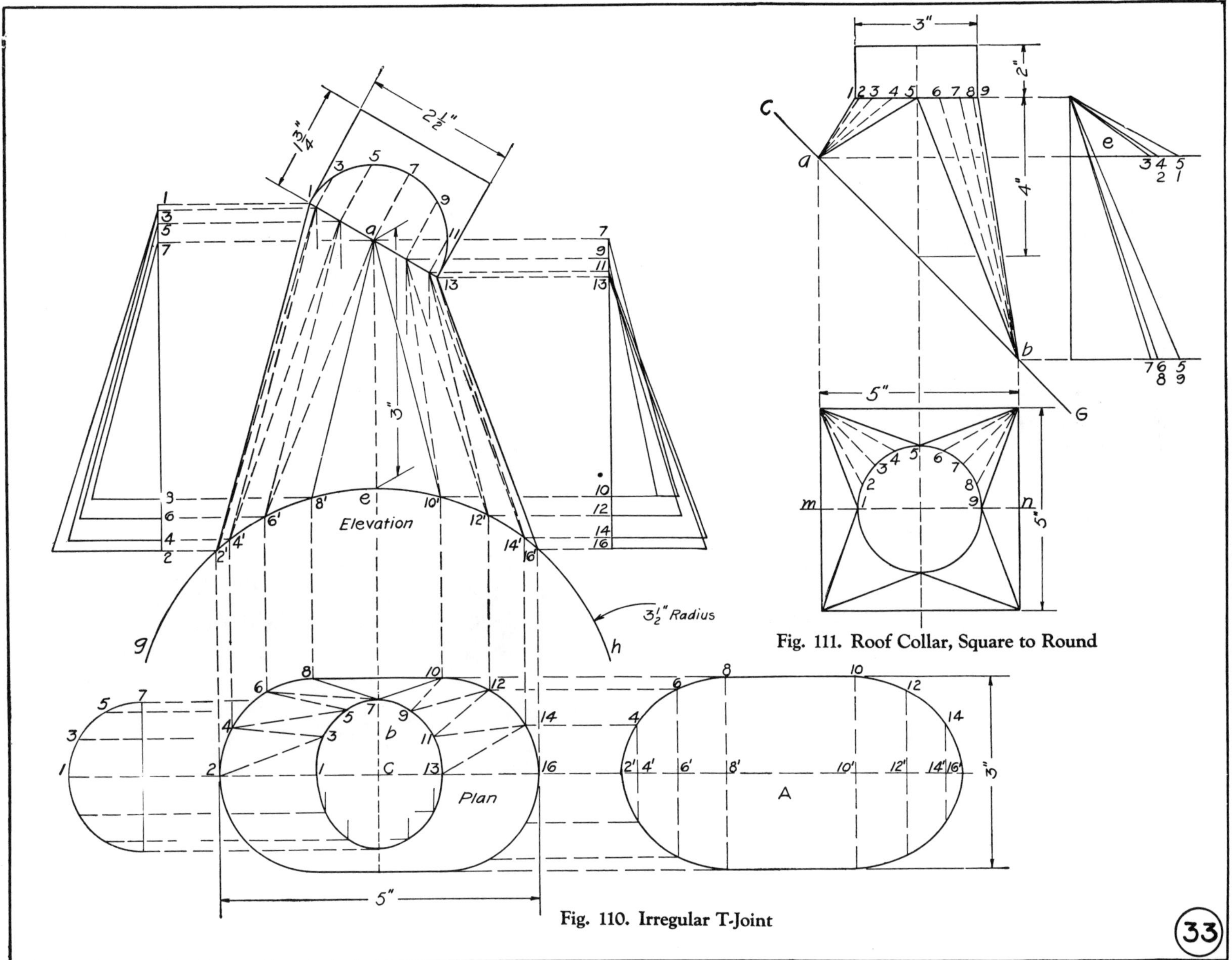

Fig. 110. Irregular T-Joint

Fig. 111. Roof Collar, Square to Round

33

shows the elevation, half plan and pattern for an oblique cone, and an inspection of its radial lines will show that they are of unequal length. The development cannot be made by the method applied to radial solids, altho the process is a combination of that method and triangulation.

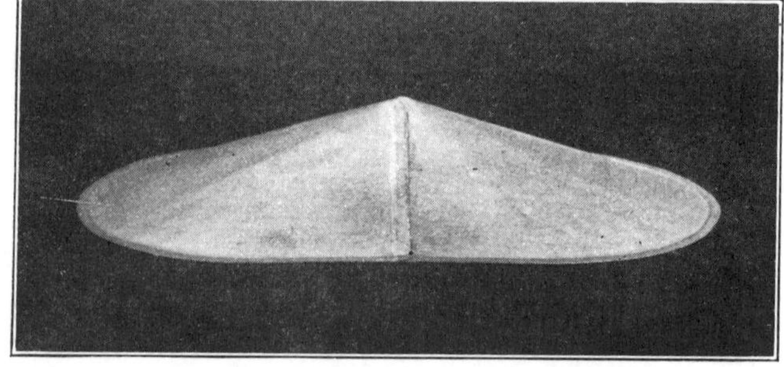

Problem 75.   Scalene or Oblique Cone.

Let *AHB* represent the elevation, and *BFH* the half plan, drawn for convenience so that the line *HB* serves as both the base line of the elevation and the center line of the plan. From the vertex *A*, draw a vertical line intersecting the base line *HB* at *C*. Divide the half circle *BFH* into equal parts, shown from *1* to *7*, and draw lines to the apex *C*. These lines will form the bases of a series of right-angled triangles of which *AC* is the altitude, and whose hypotenuses will show the lines in their true length. The most convenient method of constructing these right-angled triangles is to transfer the distances from *C* to the various points in the half plan, to the base line *HB*, measuring each time from the point *C*. Then, with *C* as center and radii equal to *C–2*, *C–3*, *C–4*, etc., draw arcs intersecting the base line *HB*.

Lines are drawn from each of these points to the apex *A*, thus producing the elements of the oblique cone in their true length, the method used being similar to that used in the other triangulation problems. Now, using *A* as center, with radii equal to *A–1'*, *A–2'*, *A–3'*, *A–4'*, etc., draw arcs, as shown. From any point upon the arc drawn from point *1'*, as *m*, draw the line *1–A*. Now, set the dividers equal to the spaces contained in the half circle in the plan view, and starting from the point *1–m*, step from one arc to another, as shown from *1* to *7*, after which complete the opposite half to *1–x*. A line traced thru these points, as shown, will be the required pattern. To obtain the pattern for the frustum of an oblique cone, shown by *GR HB* in the elevation, use *A* as center, and with radii equal to the various divisions on *GR*, draw arcs intersecting similar radial lines in the pattern. The curve *eab* traced thru these points of intersection completes the development.

Problem 76. Oblong, Raised Cover With Semi-Circular Ends

The principle given in the preceding problem is applicable to the development of the oblong raised cover with semi-circular ends, shown in Figure 113.

Problem 76.   Oblong, Raised Cover.

On examining the drawings, it will be seen that those parts of the surface bounded by the curved outline and sloping to the point *A'* in the plan are portions of an oblique cone. First, draw the plan view, the semi-circular ends being described from

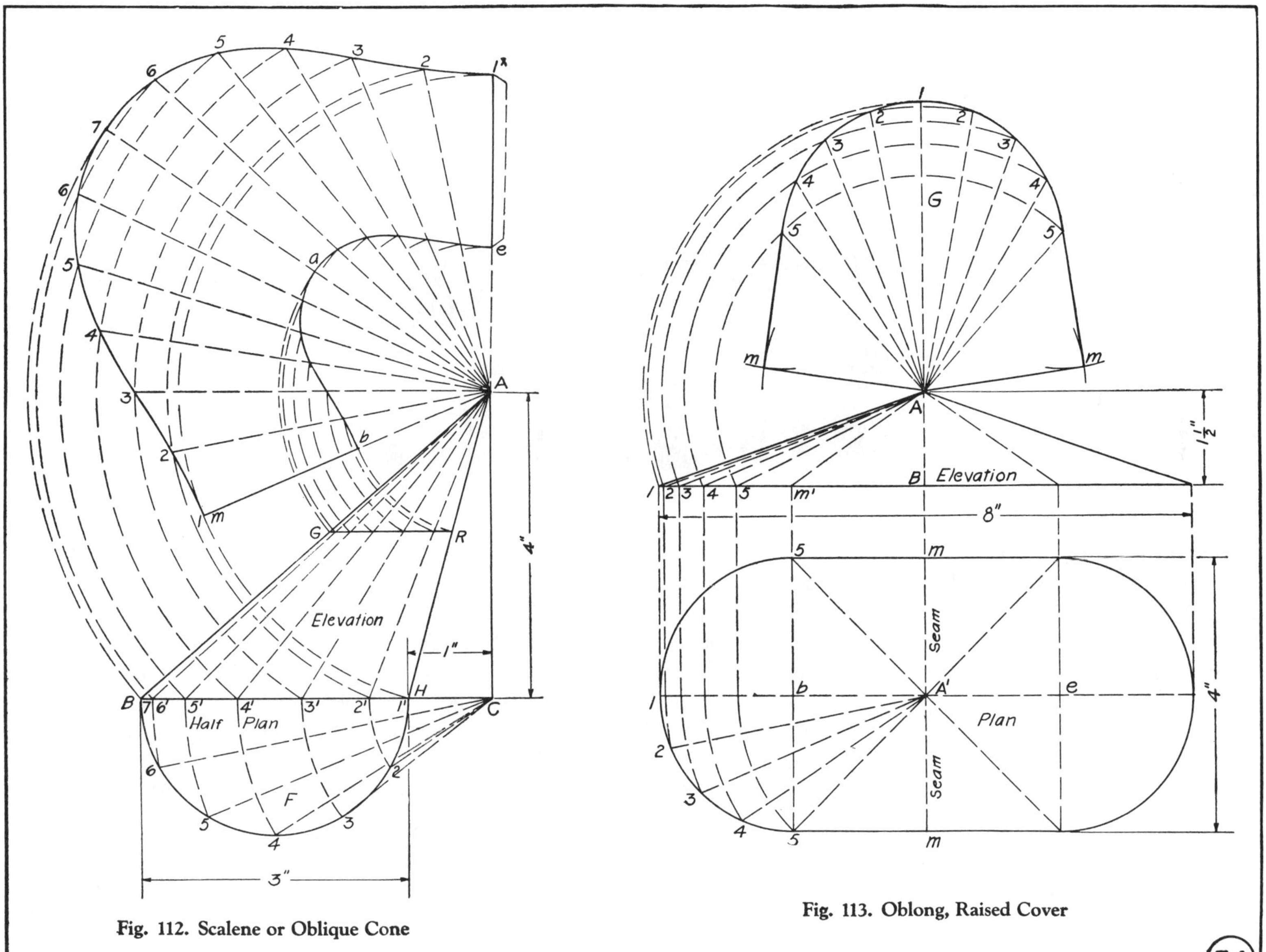

Fig. 112. Scalene or Oblique Cone

Fig. 113. Oblong, Raised Cover

34

the centers b and e. Draw a horizontal center line thru the plan, and divide one-quarter of the curved outline into a convenient number of parts, as at *1, 2, 3, 4* and *5*. From these points, draw lines to the apex A' in the plan, and let mA' represent the seam line of the cover. Next, draw the elevation, the height of the cover being shown by the center line AB. From A' in plan as center, describe arcs, respectively, from *5, 4, 3* and *2*, producing them until they reach the horizontal center line; from these intersections, draw vertical lines until they reach the base line of the cover in the elevation. Lines drawn from these points to the apex A will then represent the true lengths of similarly numbered lines in the plan. To obtain the half pattern, shown at G, use A as center, and with radii equal to $A–1$, $A–2$, $A–3$, $A–4$, $A–5$ and $A–m'$, draw arcs, as shown. From point *1* on the outer arc, draw the center line $1–A$ in the pattern. Then, with the dividers set equal to the various divisions in the quarter plan $m–1$, set off spaces on corresponding arcs in the pattern, on either side of the line $1–A$. Trace a line thru the various intersections, thus completing the half pattern for the raised cover.

Problem 77. Rectangular Raised Cover With Rounded Corners

In Figure 114 is shown an elevation and plan view of a rectangular raised cover with quarter-circle corners which is drawn to a scale of 3 inches to the foot.

The shape of the cover may, perhaps, be more accurately described as that of an oblong pyramid with rounded corners. An inspection of the drawing will show that the rounded corners are portions of a scalene cone, while the four sides are simply flat triangular surfaces.

First draw the plan view; the rounded corners being described from the centers a, e, g and c. Next, divide one of the quarter circles into a number of equal parts and from these divisions, draw lines to the apex A. Draw the elevation BFG and develop a half pattern by the method described in Problem 76. Let the line mAm in plan represent the seam line of the cover.

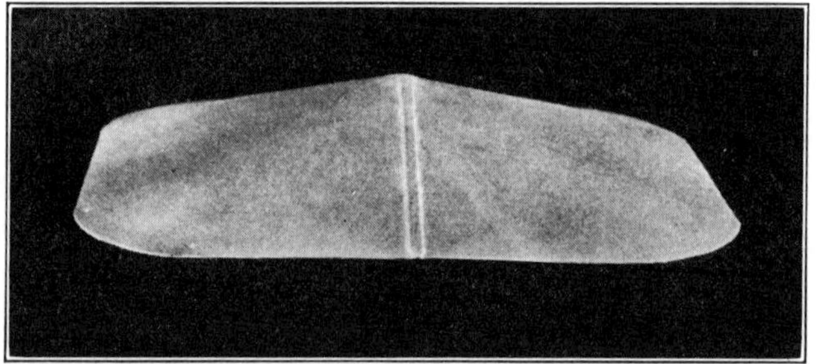

Problem 77. Rectangular Cover with Rounded Corners.

Problem 78. Six-Pointed Star

Figure 115 shows a plan view and elevation of a six-pointed star, which is drawn to a scale of 6 inches to the foot. The development of this problem does not differ materially from that in the two preceding problems. First, draw the plan view of the star, the height of which is equal to Am in the elevation. The line ab in plan is shown in its true length by the line AC in elevation. The line eb is also shown in its true length in the plan. The true length of the line ae in plan is obtained by taking the length of this line as radius, and with a as center, describe an arc intersecting the horizontal line ab at g. From point g, draw a vertical line intersecting the base line BC in the elevation, as shown at n. Draw the line An, which will show the true length of the line ae in plan.

Having found the true lengths of all lines required in the development, construct a half pattern, with the seam on the line Oe in plan. A section on the line oh in plan is shown by the outline okh and is obtained in the following manner: Extend oh

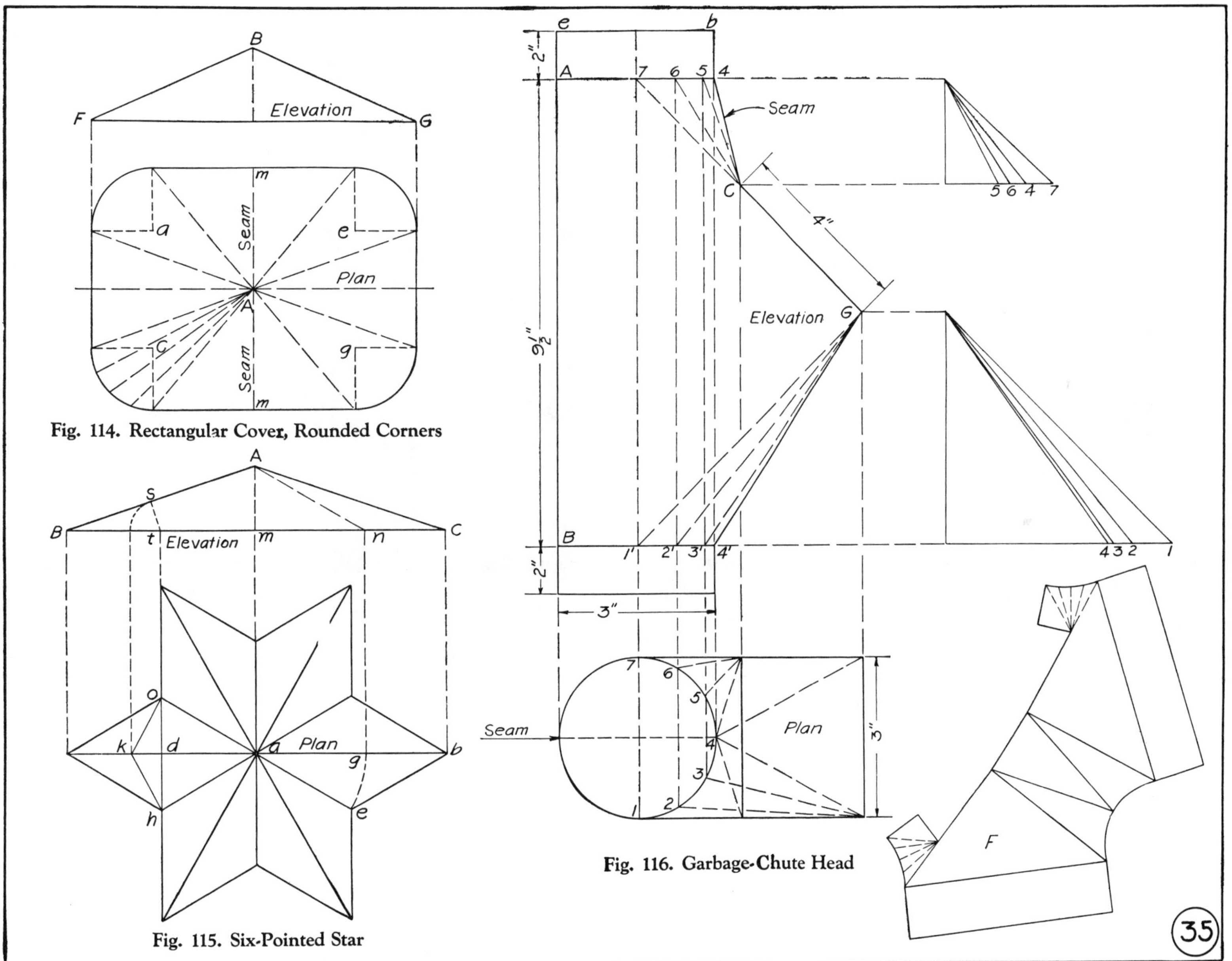

Fig. 114. Rectangular Cover, Rounded Corners

Fig. 115. Six-Pointed Star

Fig. 116. Garbage-Chute Head

35

until it intersects the base line *BC* in elevation at *t*, from which point at right angles to *AB*, draw *ts*.   Place the distance *ts* from *d* to *k* in plan.   Connect *ko* and *kh*, which will give the true profile of the desired section.

Problem 79.   Ash or Garbage Chute Head

Ash or garbage chute heads are made in many different styles dependent on the pleasure of the draftsman.   The main point to be observed in producing a design for a chute of this kind is to see that all unnecessary angles and elbows are omitted, so that no obstructions are placed in the way of the contents.

A plan and elevation of a chute head that answers these requirements is shown in Figure 116.   The body of the head is constructed from one piece of metal, with seams on *AB* and *4–C*, as shown in the drawing.   The plan is first to be drawn in accordance with the dimensions given.   First, describe the circle to represent the round pipe, and then draw the rectangle that represents the plan view of the opening in the body of the chute. The horizontal center line is now drawn thru the center of the circle and divides the rectangle into symmetrical halves.   Next, represent the outline of the round pipe in the elevation, and draw the vertical center line.   From the center point *1'* on the base line *B–4'*, draw the line *1'–G* at an angle of 45° to the base line.   Placing the triangle opposite to the position it had when

drawing the line *1'–G*, move the triangle along the line *1'–G* and locate the point *C* four inches from *G* and outside the line *4–4'* of the round pipe.   Draw the lines *C–4* and *G–4'*, completing the elevation of the body. Short collars are connected at the top and bottom, as shown by *e–b–4–A* in the elevation. Divide the outline of the round pipe in the plan into a number of equal spaces, and from these points draw vertical lines to the elevation intersecting the upper and lower base lines, as shown.

The true lengths of all lines in the plan and elevation may now be found by constructing a diagram of triangles by the method previously explained, and the pattern developed in one piece.   An outline of the completed pattern, which is reduced in size, is shown at *F*.

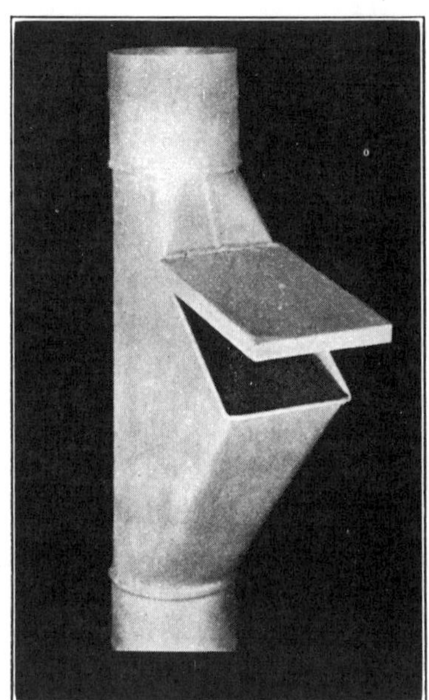

Problem 79.   Ash or Garbage
Chute Head.

CHAPTER VIII

TRIANGULATION

SIMPLIFIED METHOD

This chapter explains the simplified method of developing patterns by triangulation, in which no plan view is required, the elevation simply being used, on either ends of which are placed the semi-profiles, from which the altitudes for obtaining the true lengths of the lines in the elevation can be found.

This method can be used only when both halves of the article are alike or symmetrical. The elevation must always be drawn at a right angle to a line drawn thru the center of the plan, which divides the article into two symmetrical parts.

Problem 80. Irregular Flaring Roof Collar

The application of the simplified method of triangulation for developing patterns for irregular forms in sheet-metal work is shown in Figure 117, which is drawn one-third full size. The drawing shows the pattern and elevation for an irregular tapering roof collar that fits over a pitched roof, indicated by the line *mn*. A full view of the upper and lower bases of the collar will be exact circles. The form of the collar, altho closely resembling the frustum of a regular cone, is such that its pattern can be developed only by triangulation. First, draw the vertical

center line *4–11'* in the elevation, and thru the point *11'*, draw the slanting roof line *mn* at an angle of 30°.

Next, with a radius equal to one-half the diameter of the lower base and with the point *11'* on the roof line *mn* as center, describe the half circle representing a half profile on the lower base, as shown at *C*. The upper base and half profile, shown at *B*, is next drawn; also the lines *1–14* and *7–8*, completing the outlines of the elevation shown at *A*. Divide the half profile *B* of the upper base into a number of equal spaces, as shown from *1* to *7*, and draw a vertical line from each point at right angles to the base line *1–7*, intersecting the line *1–7*, and numbering each point to correspond with each number upon the half profile, as shown by *2', 3', 4'*, etc. In like manner, divide the half profile *C* of the lower base into the same number of equal spaces, as shown from *8* to *14*, and from each point at right angles to the base line *mn*, draw lines intersecting *mn*, and number each point of intersection to correspond with the points on the half profile *C*, as shown by *9', 10', 11'*, etc. Connect the points on the upper and lower bases by solid lines, as shown by *2'–13', 3'–12', 4'–11'*, etc. Also connect points on the base line *mn* with points on the upper base line *1–7* by dotted lines, as shown by *13'–3'*, *12'–4'*, *11'–5'*, etc., thus dividing the surface of the collar into triangles.

These solid and dotted lines in elevation show the base line of sections which will be constructed, the altitudes of which are equal to the heights in the two half profiles. If this elevation with the two half profiles were cut from cardboard or thin metal, and the half profiles *B* and *C* turned up at right angles on the lines *1–7* and *mn*, we would have a model of one-half the article, as shown in the illustration at *F*. If wires or threads were placed in position, connecting the points on the outline of the upper and lower profiles, we would at once have the true lengths of the solid and dotted lines in the elevation as if

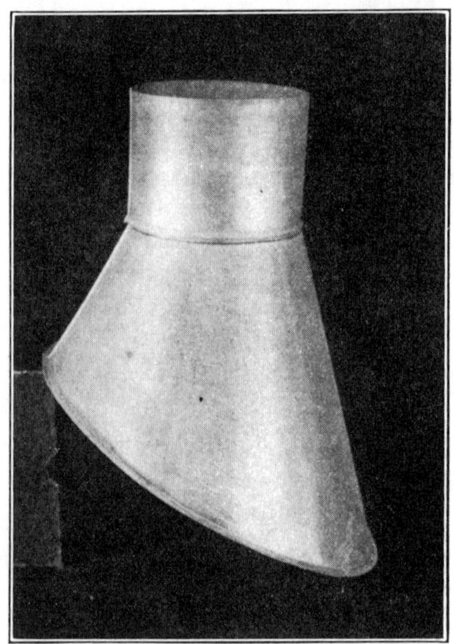

Problem 80.   Irregular Flaring Roof Collar.

measured upon the surface of the finished article. The true lengths of the solid and dotted lines are shown in the diagrams *G* and *H*, and are obtained as follows:

Take the distances in elevation of the solid lines *13'–2'*, *12'–3'*, *11'–4'*, *10'–5'*, *9'–6'*, and, measuring from *a*, place them on the horizontal line in *G*, as shown by similar numbers. From these points at right angles to the horizontal line *a–6'*, draw the vertical lines *2'–2*, *3'–3*, *4'–4*, etc., making each in length equal to

the distance of points of corresponding numbers in the half profile *B* of the upper base from the center line *1–7*, as measured upon the lines at right angles to the line *1–7*, thus obtaining the points *2*, *3*, *4*, etc.   Upon the horizontal line *a–6'*, erect a perpendicular, as *a–11*.   Upon *a–11* set off from *a* the several distances of the points in the half profile *C* from the center line *14–8* of the lower base, as indicated by the figures *11*, *10–12*, *9–13*. Now, connect these points with points on the vertical lines by means of solid lines, as shown.   Then *11–4*, *10–5*, *9–6*, etc., will represent the true lengths of similar solid lines in the elevation.

The true lengths of the dotted lines are shown in diagram *H* and are obtained in a similar manner from measurements in the elevation.   The horizontal line *14–7* is a duplicate of *a–6*, upon which set off the lengths of the several dotted lines in the elevation.   From each point draw vertical lines as before, making them equal in length to the similar lines in *G*.   The vertical line *14–11* is a duplicate of *a–11*.   Now, connect each of these points by dotted lines, as shown, thus obtaining the true lengths of similar dotted lines in the elevation.   Having obtained the true lengths of the solid and dotted lines in the elevation, the pattern is now laid out in precisely the same manner as the patterns for problems developed by the regular method of triangulation explained in the previous chapter.   To develop the full pattern, shown at *D*, first draw the center line *7–8*, making it equal in length to the line *7–8* in the elevation, which is shown in its true length.

With *8* as center and radius equal to *8–9* of the half profile *C*, describe small arcs, which intersect with another struck from *7* as center, with a radius equal to the dotted line *7–9* in diagram *H*, thus establishing the position of the two points numbered *9* in the lower edge of the pattern.   From *7* as center, with a radius equal to *7–6* of the half profile *B* of the upper base, strike

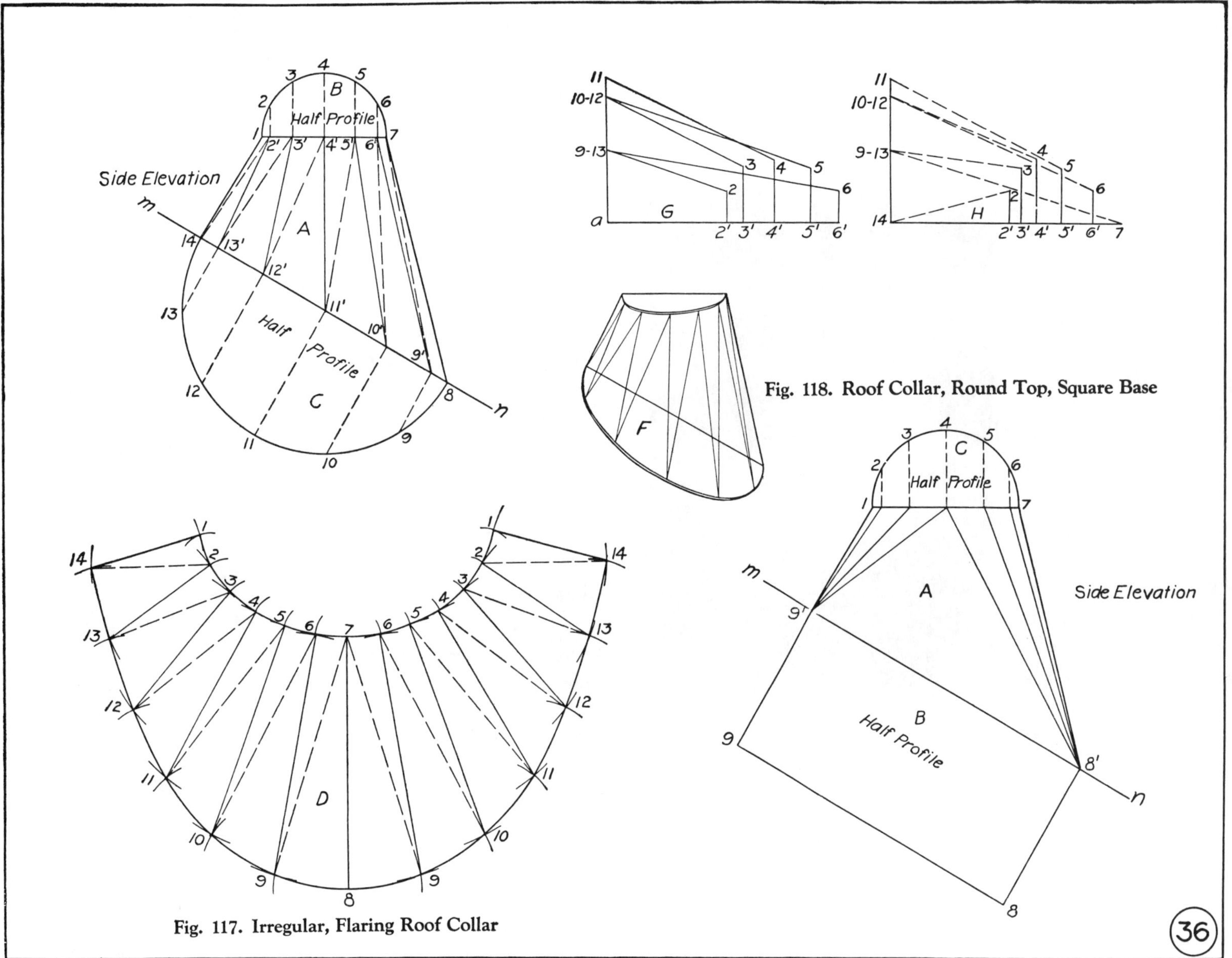

Fig. 118. Roof Collar, Round Top, Square Base

Fig. 117. Irregular, Flaring Roof Collar

small arcs, which intersect with other arcs struck from points *9* as centers, with a radius equal to the solid line *9–6* in diagram *G*, thus locating the position of point *6* in the upper edge of the pattern. Continue using the true lengths of the solid and dotted lines in the diagrams *G* and *H*, in connection with the lengths of the spaces in the half profiles *B* and *C*, to develop the upper and lower lines of the pattern, using each combination alternately until the pattern is completed. As each point on the pattern is located, it should be numbered, and the solid and dotted lines may be drawn across the pattern, as shown. These lines are not necessary, as each point is simply used as a center from which to find the next point beyond. No matter what shape the profiles of the ends of the article may be, the above method of development can be used in every case where the two halves of the article are symmetrical.

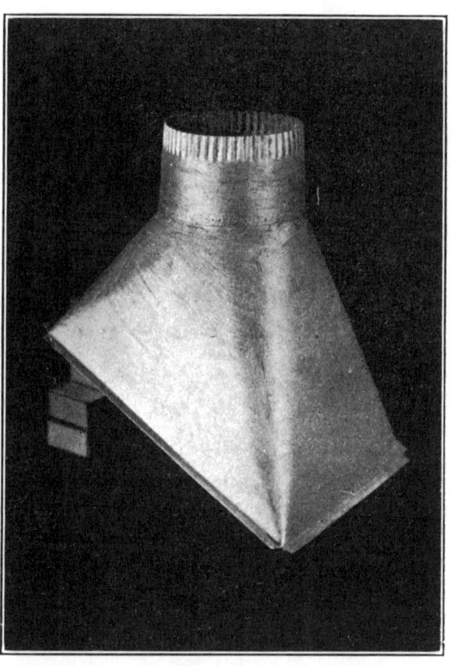

Problem 81.   Roof Collar, Round Top and Square Base.

Problem 81. Roof Collar Having Round Top and Square Base

In Figure 118, *1–7–8'–9'* is the elevation of a roof collar having a round top and square base to fit over a slanting roof, indicated by the line *mn*. *A* shows the elevation, *C* the half profile of the top, and *B* the half profile of the base. The drawing is made to a scale of 4 inches to the foot. Draw the elevation and half profiles of the upper and lower base.

Find the true lengths of the solid lines in the elevation by the method explained in the previous problem, and develop a full pattern, placing the seam on the long side of the collar, as shown by the line *7–8* in the elevation.

Problem 82. Three-Piece Offset Fitting

Figure 119 shows the method employed when offset pieces and transition elbows are developed in three pieces, and the principles given in Problem 80 can be applied to any tapering transition piece, no matter what profile either end may have.

Figure 119 shows the elevation of a three-piece tapering offset fitting, a portion of a round pipe joining it above and below. *B* is the upper arm with half profile, shown by *A*, and *G* the lower arm, the half profile of which is shown at *F*. The miter lines *1'–7'* and *14'–8'* are not found by bisecting the exterior angles, but may be established at pleasure.

The half profiles *A* and *F* are divided into the same number of equal spaces and from these points draw vertical lines parallel to the upper and lower arms until they intersect the miter lines *1'–7'* and *14'–8'*. A one-half pattern for the upper arm *B* and the lower arm *G* is shown at *H* and *D*. These are obtained by the parallel method as explained in Problem 1 on the development for a two-piece elbow in a round pipe. The pattern for the middle section *C* forms a transition piece and is developed by triangulation as follows:

Draw solid lines in *C*, connecting the points on the miter lines, as *2'–13'*, *3'–12'*, *4'–11'*, etc.; then draw diagonal dotted lines, as shown. Next, obtain the true lengths of the solid and dotted

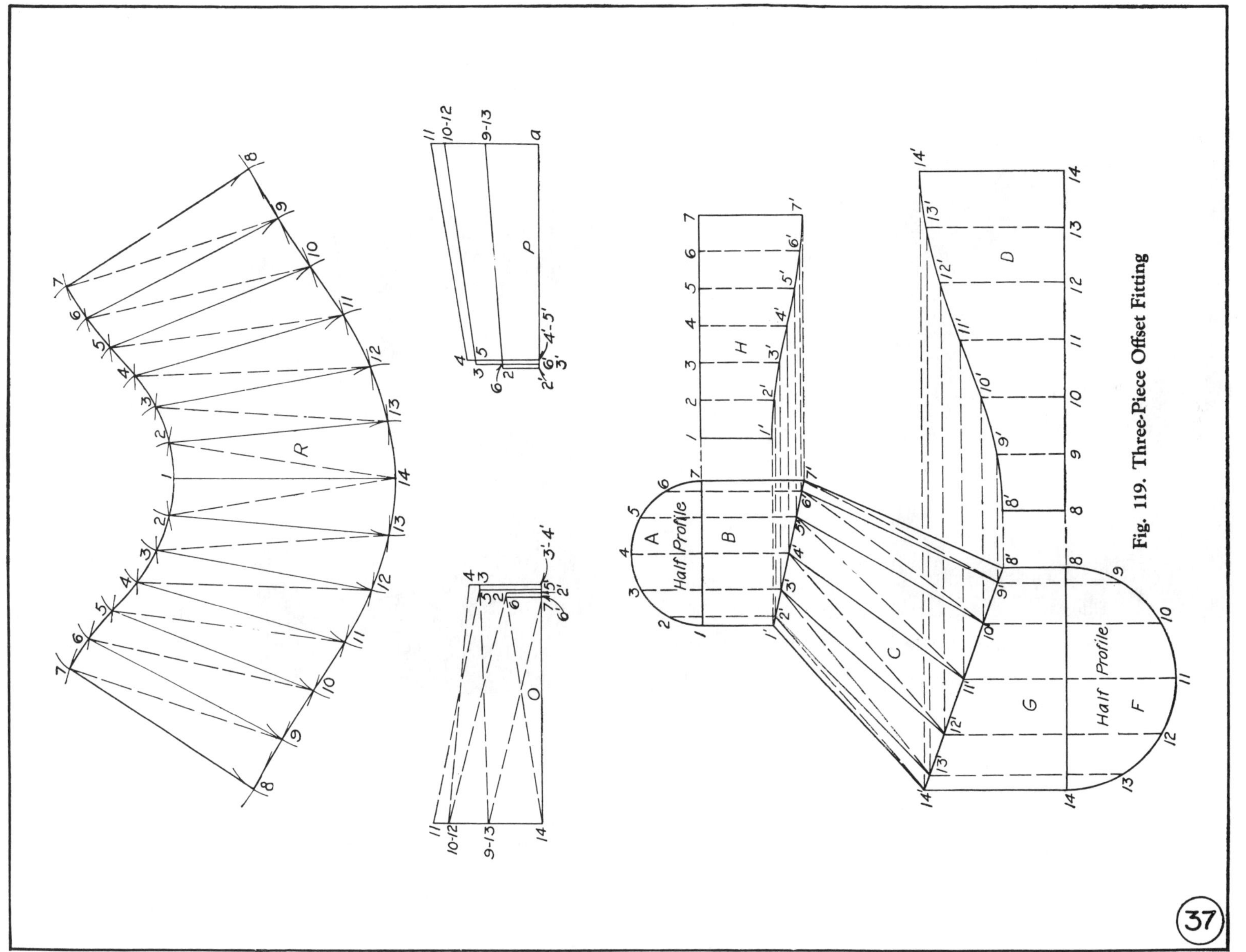

Fig. 119. Three-Piece Offset Fitting

37

lines in *C* by taking those distances and placing them on the horizontal lines in diagrams *O* and *P*, as shown by similar numbers, and from these points erect vertical lines, making them

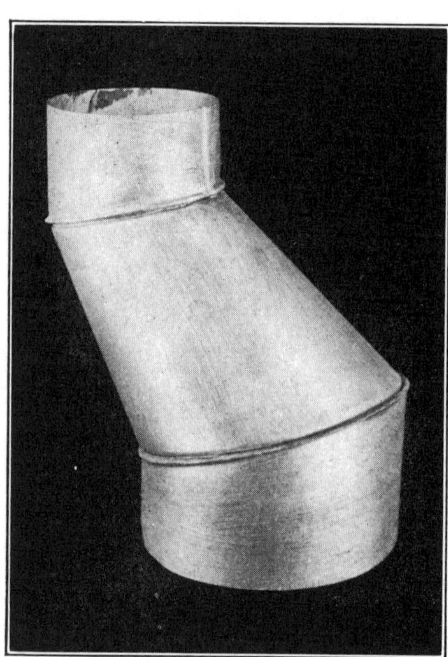

Problem 82.   Three-Piece Offset Fitting.

equal in length to similarly numbered lines in the half profile *A*, which are measured from the center line *1–7*. Upon the end of the horizontal line in *O* and *P*, erect a perpendicular, as *14–11* and *A–11*. Upon these lines set off from *14* and *A* the several distances of the points in the half profile *F*, measuring in each case from the center line *14–8* of the lower base, as indicated by the figures *11, 10–12, 9–13.* Now, connect these points with points *1, 2,*

3, 4, etc., on the vertical lines previously drawn, by means of solid and dotted lines, as shown. These lines will then represent the true lengths of similarly numbered solid and dotted lines on middle section *C* in the elevation. After the true lengths have been obtained, the pattern for *C* is developed in exactly the same manner as explained in connection with developing the pattern *D* in Figure 117; the only difference is that the spaces on the upper and lower edge of the full pattern shown at *R* are not equal in length and the various distances are taken

from the miter cuts on the half patterns *H* and *D*. Take the distance *1'–14'* in *C*, which shows its true length, and place it on the center line of the pattern, as shown by *1–14* in *R*. Using *1'–2'* in half pattern *H* as radius, and *1* in pattern *R* as center, describe arcs *2* and *2*, which intersect by arcs struck from *14* as center with a radius equal to the dotted line *14–2* in diagram *O*.

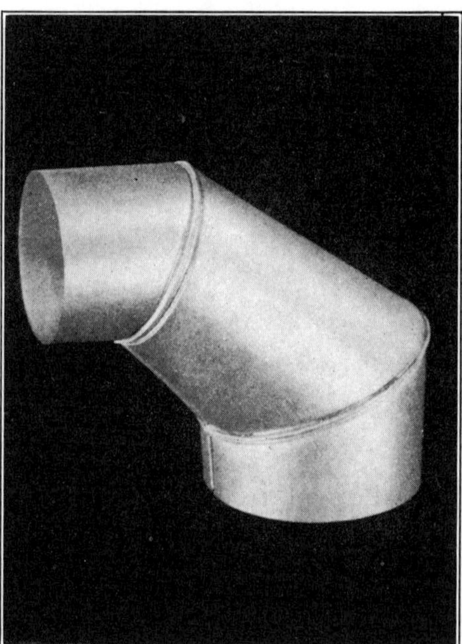

Problem 83.   Three-Piece Reducing Elbow.

Now, with *14'–13'* in half pattern *D* as radius and *14* in *R* as center, describe the arcs *13* and *13*, which intersect by arcs, struck from *2* and *2* as centers, with a radius equal to the solid line *2–13* in diagram *P*. Proceed in this manner, using alternately as radius, first, the divisions in the miter cut in pattern *H*, then the true lengths of the dotted lines in diagram *O'*; the divisions in the miter cut in pattern *D*, then the true lengths of the solid lines in diagram *P*. The edge lines *7–8* in the pattern are shown in their true lengths by the line *7'–8'* in the elevation.

Problem 83.   Three-Piece Reducing Elbow, Round to Round

Figure 120 presents the elevation, plan and half patterns for a three-piece reducing elbow, and shows the method of proce-

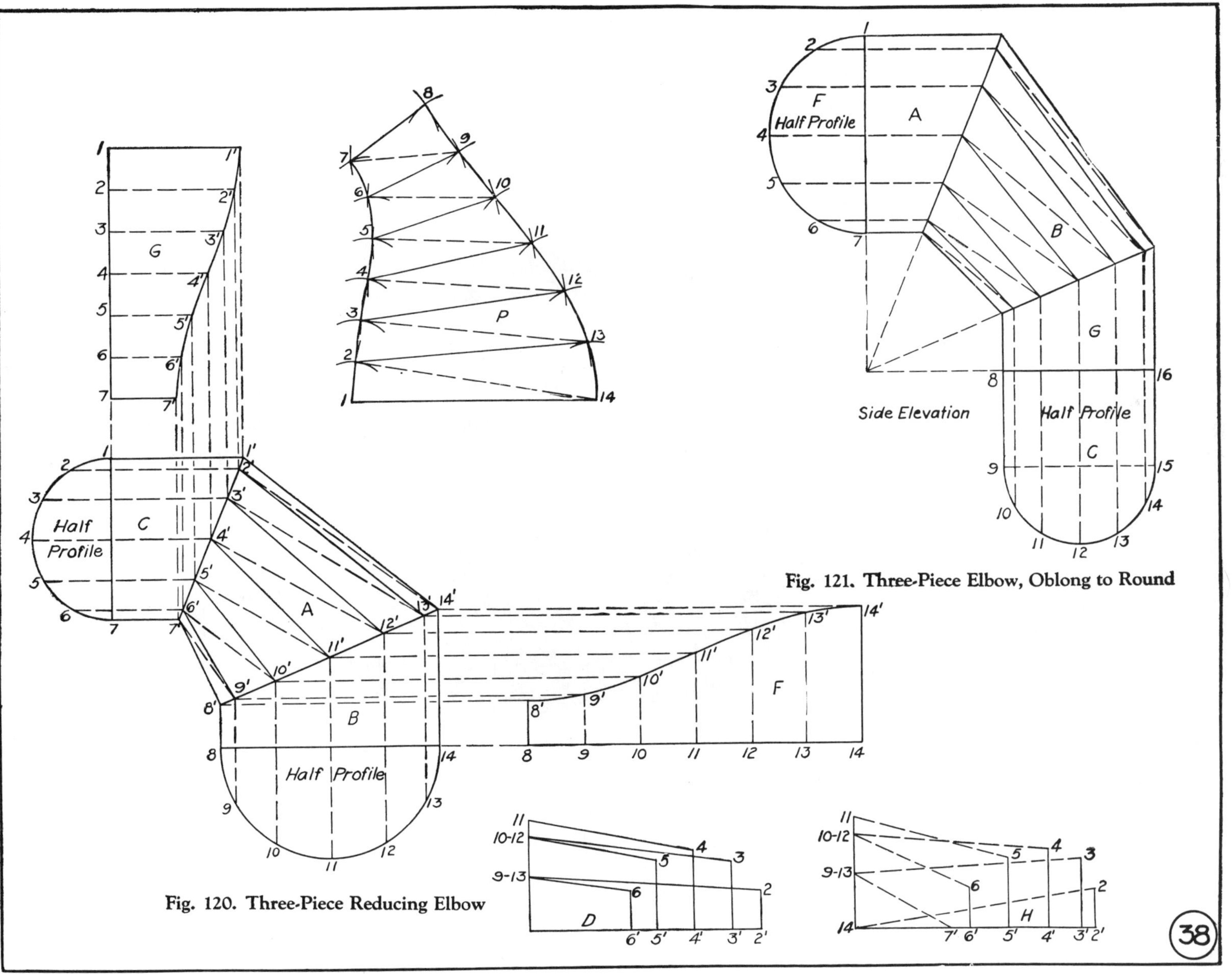

Fig. 121. Three-Piece Elbow, Oblong to Round

Fig. 120. Three-Piece Reducing Elbow

38

dure when the halves are symmetrical, and the patterns are developed without the aid of a plan. In this problem, the upper and lower pieces, shown by *C* and *B*, are developed by the parallel-line method, and the middle piece *A* by triangulation. First, draw the elevation and place a half profile at the end of *C* and the half profile at the end of *B*. Divide both profiles into an equal number of spaces, as shown by the figure *1* to *7* and *8* to *14*.

From the divisions *8* to *14* in the lower half profile, draw vertical lines until they intersect the miter line from *14'* to *8'*, and develop the half pattern for *B*, as shown at *F*. In corresponding manner, from the points *1* to *7* in the half profile of the upper piece *C*, draw horizontal lines which intersect the miter line from points *1'* to *7'*, and develop the pattern for the upper arm *C* in the usual manner, as shown at *G*. Now, connect the points on the miter lines by means of solid and dotted lines. The solid and dotted lines in middle piece *A* will then represent the bases of sections to be constructed, whose altitudes are equal to the various heights in the half profiles. The true lengths of the solid and dotted lines in *C* are shown by correspondingly numbered lines in diagrams *D* and *H*, and are obtained by the method described in Problem 82. The half pattern for the middle piece, shown at *P*, is also laid out as described in the previous problem. The divisions between *14* and *8* in the half pattern *P* are obtained from the divisions along the miter cut in the half pattern *F*, and the divisions from *1* to *7* in half pattern *P* are obtained from the divisions along the miter cut in the half pattern for *C*, shown at *G*.

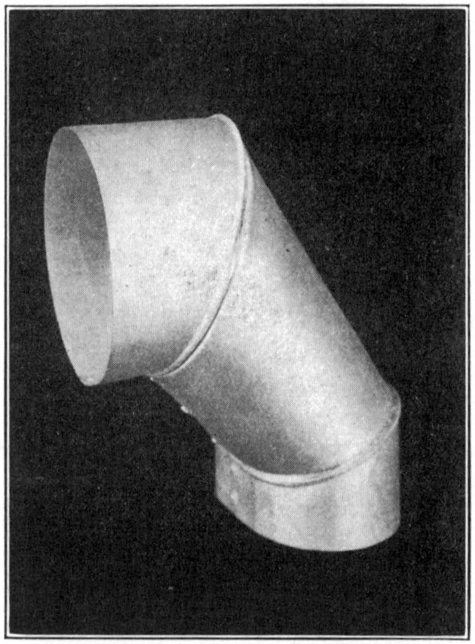

Problem 84. Three-Piece Elbow, Oblong to Round.

Problem 84. Three-Piece Elbow, Oblong to Round

In Figure 121 is shown the side elevation of a three-piece transition elbow, oblong to round in form. A half profile of the round horizontal arm *A* is shown at *F*, while the half profile of the oblong vertical arm *G* is shown at *C*. The middle section *B* is a transitive piece necessary to form a connection between the upper and lower arms, and is developed by triangulation. The patterns for both the upper arm *A* and lower arm *G* are developed by the parallel-line method, and the conditions given in this problem are essentially the same as those of Problem 83. It makes no difference whether the transition piece is shaped as shown in Figure 120, or as shown in this problem, the same methods apply in each case.

Problem 85. Three-Piece Transition Elbow, Round to Square

The principles of the simplified method of triangulation may be applied to any three-piece elbow whose halves are symmetrical, without respect to the shape of the end pieces or the angle of the elbow. Should an instance arise where the halves as viewed in plan are not alike, a plan view must necessarily be used. Figure 122 shows the side elevation of a three-piece elbow, of which the lower arm *A* is round, and the upper arm *C* is square. The middle section *B* is a transition piece from round to square, and is developed by triangulation.

Draw the elevation and place the profile *G* on the end of *C* and

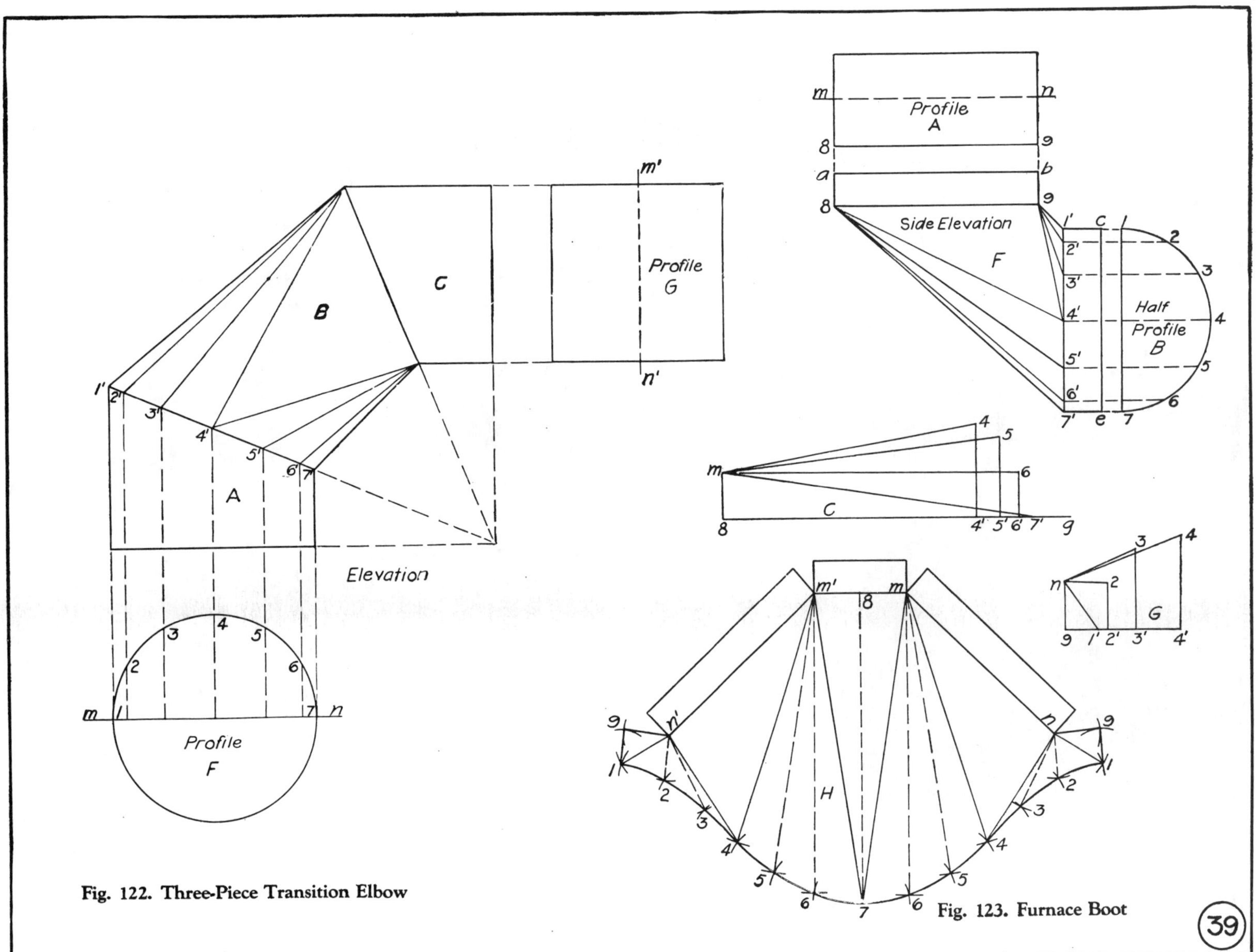

Fig. 122. Three-Piece Transition Elbow

Fig. 123. Furnace Boot

39

profile F in its proper position below A. Develop the patterns for the end pieces A and C by the parallel-line method, and the pattern for the transition piece B by the simplified method of triangulation, as described in the preceding problems in this chapter.

Problem 86. Furnace Boot, Round to Rectangular

Figure 123 shows the method of developing the pattern for an angular furnace boot, round to rectangular. First, draw the side elevation of the boot, as shown, and above the upper end place the profile of the rectangular pipe, as shown at A. In its proper position on the lower end, draw the semi-circle representing the half profile of the round pipe, shown at B. Divide the semi-circle into a number of equal parts, as shown from 1 to 7. From these points on the half profile, draw lines at right angles to the center line 1–7 intersecting the line $1'$–$7'$ at $2'$, $3'$, $4'$, $5'$ and $6'$.

From $6'$, $5'$ and $4'$ draw lines to the upper corner 8, and from $4'$, $3'$ and $2'$ draw lines to corner 9. The next step is to find the true lengths of the various lines in the side elevation, and it will be necessary to construct the dia-

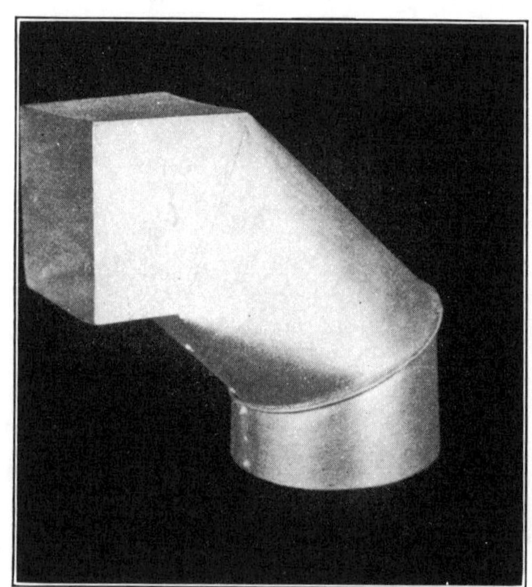

Problem 85.   Three-Piece Transition Elbow Round to Square.

grams, shown at C and G, by the method explained in the previous problems. To find the true length of the line $4'$–8 in the side elevation, take this distance and place it on the horizontal line 8–g, as shown from 8 to $4'$ in diagram C. From 8 and $4'$ draw perpendicular lines, making 8–m equal to m–8 in profile A, and $4'$–4 equal to the distance from the line 1–7 to point 4 in the half profile B. Then m–4 in dia-

Problem 86.   Furnace Boot, Round to Rectangular.

gram C is the true length of the line 8–$4'$ in the side elevation, and all the true lengths are obtained in the same manner. The first step to construct the pattern H is to draw the center line 8–7 equal in length to 8–$7'$ in the elevation, which is shown in its true length. At right angles to 7–8 draw the line m–m' equal to the width of profile A, and draw lines from m and m' to 7 in pattern H.

With radii equal to m–6, m–5 and m–4 in diagram C, and with m and m' as centers, describe the small arcs, 6, 5 and 4. Now set the dividers equal to the spaces contained in the half profile

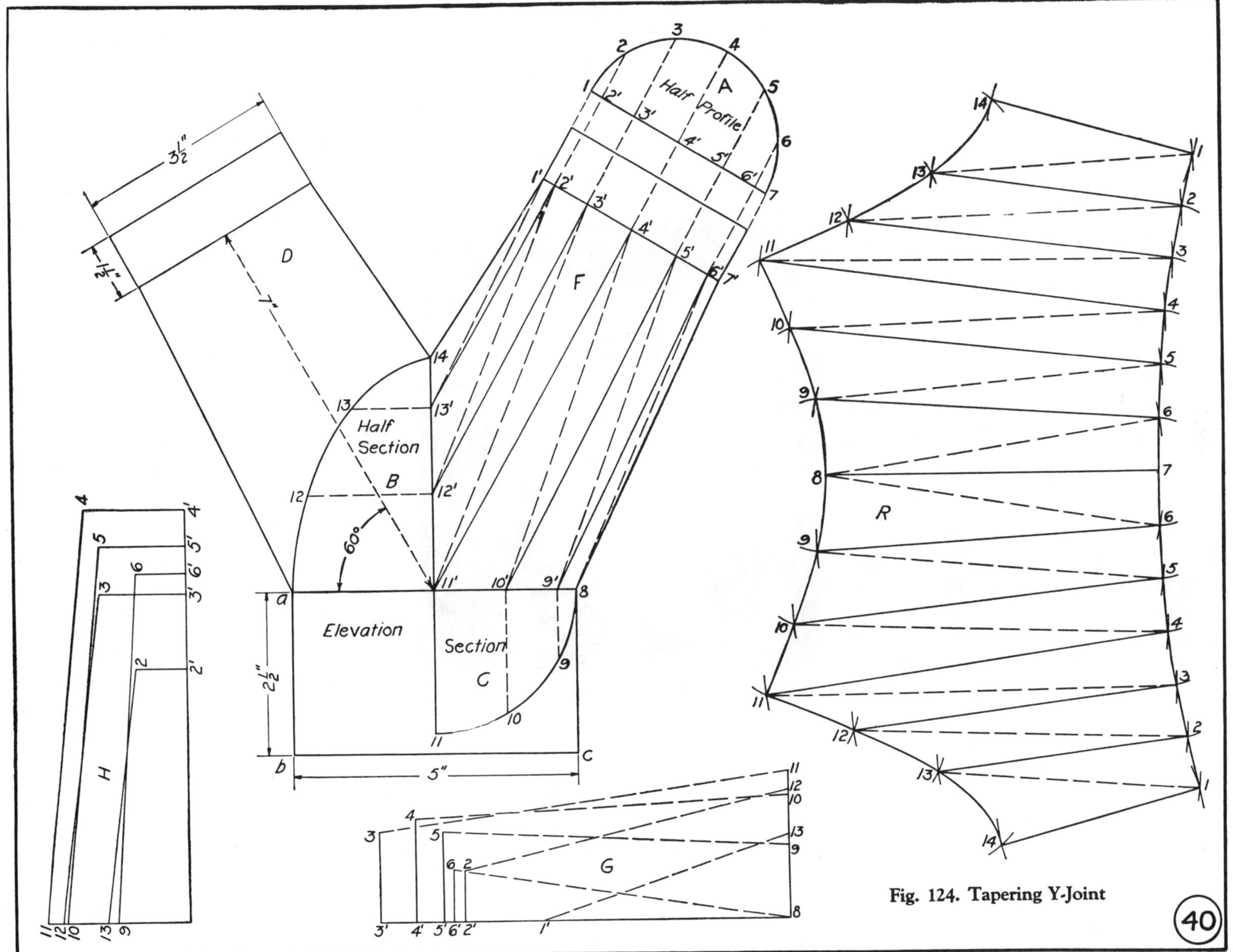

Fig. 124. Tapering Y-Joint

40

B in the elevation, and, starting at *7* in pattern *H*, step from *6*, one arc to another, and draw lines from *m* and *m'* to *4*. With *mn* in profile *A* as radius, and with *m* and *m'* in pattern *H* as centers, describe the arcs *n* and *n'*, which are intersected by arcs struck from *4* as center and *n–4* in diagram *G* as radius. Now with *n* and *n'* as centers, and with *n–3*, *n–2* and *n–1* in diagram *G* as radii, describe the arcs *3*, *2* and *1*. Set the dividers equal to the spaces *4–3*, *3–2* and *2–1* in the semi-circular profile, and, starting at point *4* in the pattern, step to arcs *3*, *2* and *1*, respectively. Draw a line from *1* to *n* and *n* to *4*. Now, with radius equal to *9–1'* in the elevation, which shows its true length, and *1* in pattern as center, describe the arc *9*, which is intersected by an arc struck from *n* as center and *9–n* in the rectangular profile *A* as radius. Complete the pattern by tracing a curved line thru the various points, and to the upper edge add the vertical collar, shown by *a–b–8–9*, in the side elevation. The lower collar, shown by *1'–C–7'–e*, is simply a straight piece of round pipe, for which a pattern is not required.

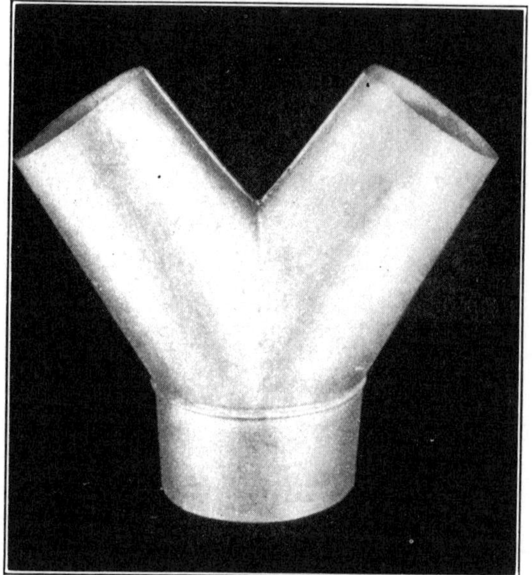

Problem 87. Two-Piece Tapering Y-Joint.

Problem 87. Two-Piece Tapering Y-Joint

There are many problems arising in the course of the sheet-metal worker's experience where it is necessary to take from a leader or main pipe, several branches, each of which shall run in a different direction, and the fitting should be so constructed as to provide an easy flow for the contents of the pipe.

Figure 124 illustrates the principles for developing a Y-joint, consisting of two tapering pipes joining a larger pipe at an angle, and is a fitting that is generally recognized by sheet-metal workers as well adapted to work of this kind. The elevation of the larger pipe is shown by *a*, *b*, *c*, *8*, and is simply a straight piece of pipe with which the branches of the Y are to be joined at its upper end, shown by the line *a–8*. Bisect the line *a–8*, and from this point, as at *11'*, draw the line *11'–4* at the given angle to serve as the center line of prong *F*. At any convenient point on this line, as *4'*, erect the perpendicular *1'–7'*, making it equal in length to the diameter of the small end of the tapering branch. Next, draw the vertical center line *14–11'*, and connect *14–1'* and *7'–8*, completing the elevation of branch *F*.

The center line *14–11'* represents the seam line or miter line that separates the two branches of the fitting, and the draftsman must select an outline that shall be adapted for a full view of a section on this line. The usual practice is to draw an irregular curve from *14* to *a*; then, *14–a–11'* will represent a half section on *14–11'*. With *11'* as center and *11'–8* as radius, describe the quarter circle *8–11*, which will represent a quarter section of the large pipe, as shown at *C*. Now, draw the half profile *A* and divide the semi-circle into a number of equal parts, as shown from *1* to *7*. From these points and at right angles to the center line *1–7*, draw lines which intersect the line *1'–7'*, shown by *1*, *2'*, *3'*, *4'*, etc., on the upper end of branch. Since a number of points are to be located on the miter line and the lower base of the branch, these points should be equal in number to those on the upper end.

An inspection of the drawing will show that of the six spaces required for the lower end of branch *F*, three of them are located on the outline of the quarter section, shown by the points *9* and *10* at *C*, and the remaining three on the outline of the half section on the miter line, shown by the points *12* and *13*. From the points *12* and *13*, draw horizontal lines, which intersect the miter line at *12'* and *13'*, and from points *9* and *10*, draw vertical lines to the base where they are designated by the numbers *9'–10'*. Connect the various points on the upper and lower end of the branch by solid and dotted lines drawn in the usual manner.

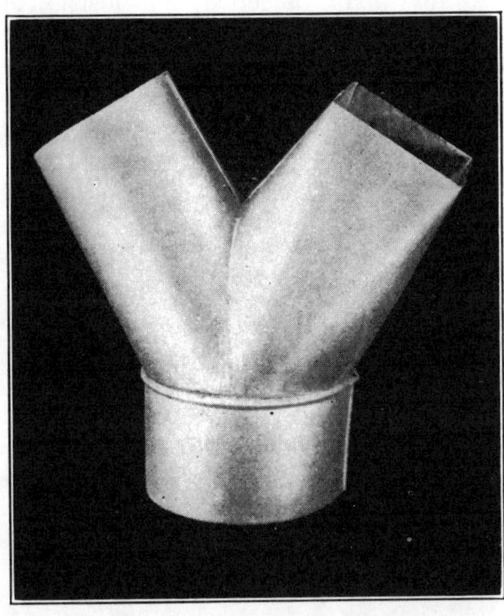

Problem 88.   Irregular Two-Branch Y-Joint.

The true lengths of these lines are shown in diagrams *G* and *H* and are obtained by the method previously shown for such cases. For example: To obtain the true length of the dotted line *11'–3'* in *F*, place the length of that line on the horizontal line in diagram *G*, shown from *8* to *3'*; erect a perpendicular on each end of this line, making *3'–3* equal to *3'–3* in half profile *A*, and *8–11* equal to *11'–11* in half section *C*; draw a line from *11* to *3* in diagram *G*, which is the true length of the dotted line *11'–3'* in *F*. The true lengths of all the solid and dotted lines are found

in a similar manner.   As both branches *F* and *D* are similar, the pattern for one branch only is required.

Placing the seam on the short side of the branch, as *1'–14* in *F*, the pattern is developed as follows: Draw the center line *7–8* in pattern *R* equal in length to *7'–8* in the elevation.   With *7–6* in half profile *A* as radius and *7* in *R* as center, describe the arc *6*, which intersect by an arc struck from *8* as center and *8–6* in diagram *G* as radius.   Then, with *8–9* in quarter section *C* as radius and *8* in *R* as center, describe the arc *9*, which intersect by an arc struck from *6* as center and *6–9* in diagram *H* as radius.

Proceed in this manner until the line *4–11* in the pattern is obtained.   The divisions *12, 13, 14* in the pattern are then obtained from half section *B*.   Then, using *4–3* in half profile *A* as radius and *4* in pattern *R* as center, describe the arc *3*, which intersect by its proper radius found in *G*.   With *11* in the pattern as center, and *a–12* in half section *B* as radius, describe the arc *12*, which intersect by an arc struck from *3* as center, and *3–12* in diagram *H* as radius; then use the spaces in the half sections *A* and *B*, and the true lengths in diagrams *G* and *H*, until the line *1–14* is drawn, which is shown in its true length from *1'* to *14* in *F*.   A line traced thru the various points completes the pattern.   Laps should be allowed for seaming together, as well as for seaming to the collars which are placed at each opening.

Problem 88.   Irregular Two-Branch Y-Joint

Figure 125 shows the elevation and half sections of an irregular two-branch Y-joint, which is drawn to a scale of 3 inches to the foot.   The main pipe, shown by *mneh*, is round in form, the opening in the upper end of branch *B* is square, and the opening in branch *A* is round.

The half circle *F* represents the half section on the base line *mn*, *G* the half section on the line *bd*, and *C* the half section of the square pipe on the line *ug*.   *H* in the side elevation shows the half section on the miter line *7–a*.   Draw the elevation and sections, as shown, and develop the pattern for each branch by the method explained in the previous problem.

Problem 89.   Two-Branch Y-Joint, Round to Round

Figure 126.   This figure is drawn to a scale of 3 inches to the foot, and shows the elevation and sections of a two-branch tapering Y similar in form to Figure 124.   The branches *A* and *B* are mitered to vertical pipes *F* and *G*.   Draw the elevation of the prongs; also the half profiles of the vertical pipes and the quarter section *C* of the large pipe. The height of the miter line between the two prongs at *14–a* is made equal to the semi-diameter *ma* of the large pipe. In other words, the true half section *R* on the miter line *14–a* constitutes a quarter circle with a radius equal to one-half the diameter of the large pipe. Having drawn the half profile *H* and the sections *R* and *C*, find the true lengths of the solid and dotted lines and develop

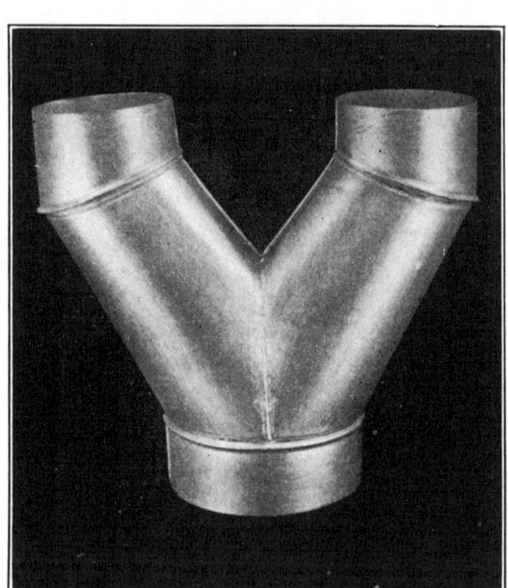

Problem 89.   Two-Branch Y-Joint, Round to Round.

the pattern for branch *A* in the same manner as described in connection with branch *F* in Figure 124.   The divisions between *1'* and *7'* on the upper edge of the pattern are obtained from the divisions along the miter cut of the pattern for the vertical pipe *G*, which is developed by the parallel-line method, the stretch-out line being placed in the position shown by the line *be*.

Problem 90.   Two-Branch Y-Joint, Round to Square

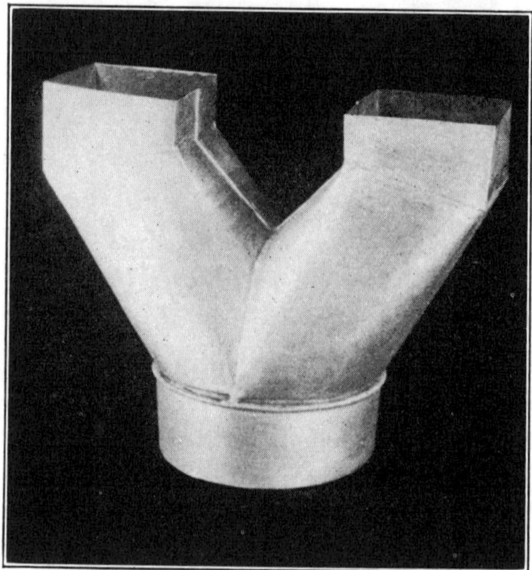

Problem 90.   Two-Branch Y-Joint, Round to Square.

The problem presented in Figure 127 is similar to that described in the previous problem, with the difference that the upper end of both branches are mitered to vertical pipes *A* and *H*, which are square in form; the half profiles are shown at *B* and *R*.   The true half section on the miter line *7–a* is shown at *F*, and is obtained in the same manner as in Figure 124.   The quarter circle *n–4* represents a quarter section of the round pipe, shown at *G*.   Figure 127 is drawn one-third full size, and, when enlarging this and the various problems in this chapter, more spaces should be used in dividing the profiles.

Draw the elevation, half profiles and sections, as shown, and develop the patterns for branch *C* and the vertical square pipe

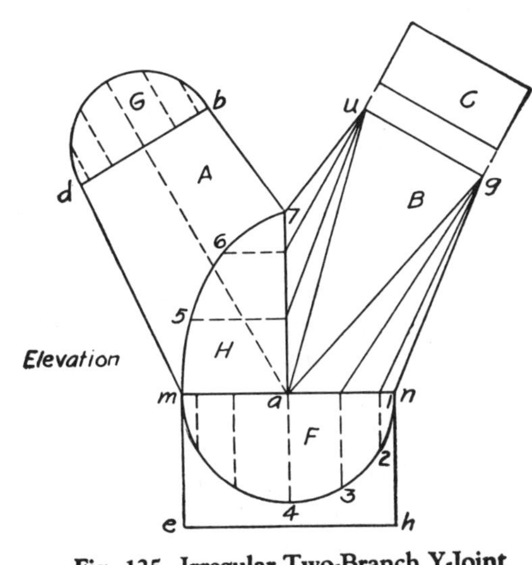

Fig. 125. Irregular Two-Branch Y-Joint

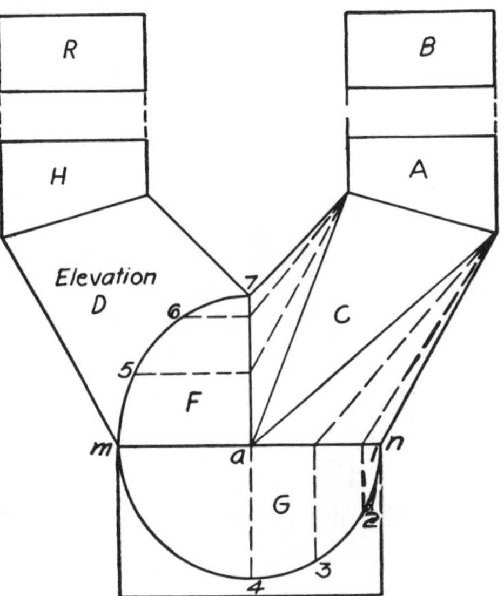

Fig. 127. Y-Joint, Round to Square

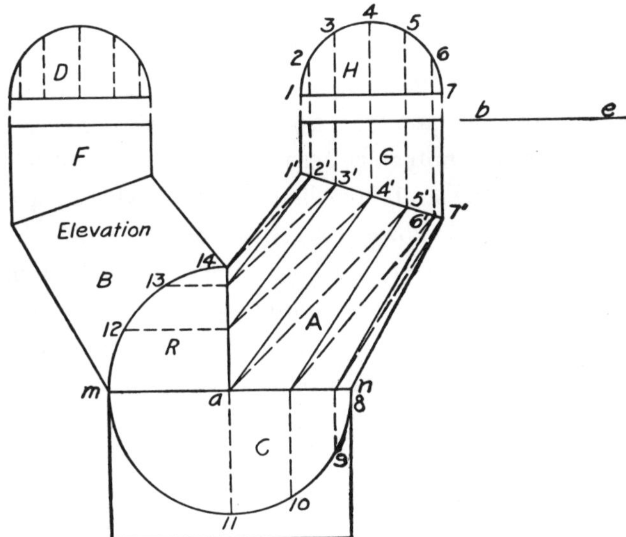

Fig. 126. Two-Branch Y, Round to Round

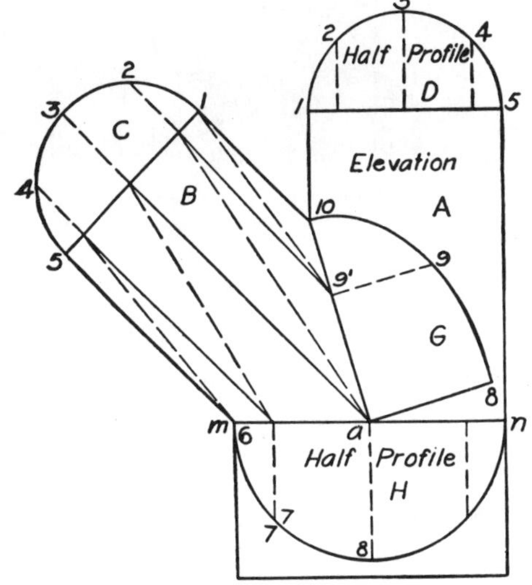

Fig. 128. Irregular Two-Branch Fitting

41

A. As both branches *C* and *D* are similar, only one pattern is required.

Problem 91. Irregular Two-Branch Fitting, Round to Round

Figure 128 shows the side view and sections of a two-branch fitting which is drawn to a scale of 3 inches to the foot. The main pipe and branches are round in form, the half profiles of which are shown by the semi-circles *C*, *D* and *H*. Branch *A* is in a vertical position, while branch *B* is inclined at an angle of 45° to the base line *mn*. The miter line between the branches *A* and *B* is obtained by drawing a line from the intersection of the

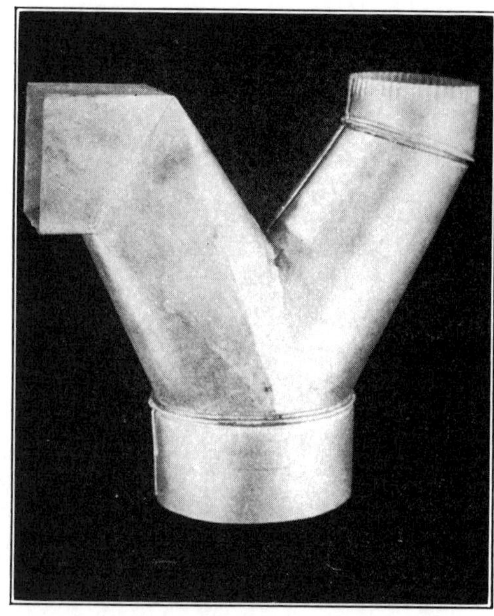

Problem 92.   Irregular Fitting, Round to Square.

two branches at point *10*, to the center of the upper base of the large pipe, shown at *a*. The half section on the miter line *10–a* is shown at *G*, and is drawn as follows:

From point *a* at the lower end of the miter line, and at right angles to *10–a*, draw the line *a–8*, making it equal in length to the semi-diameter of the large pipe, as shown by *a–8* in half profile *H*. Connect points *10* and *8* by means of an irregular curve. Then, *10–8–a* will represent a half section on miter line *10–a*. Draw the

elevation and sections, as shown, and develop separate patterns for branches *A* and *B* by the method described in Problem 87.

Problem 92. Irregular Fitting, Round to Square, Mitering with Vertical and Horizontal Connections

In figure 129, *ABRCF* represents the side elevation of an irregular Y fitting, the base *R* being round, the branch *C* mitering with the horizontal square pipe *F*, and branch *B* mitering with the vertical round pipe *A*. The circle *D* represents the section on the line *ab*, *H* the section of the square pipe on the line *Ce*, and the semi-circle *R* the half section of the large pipe on the line *6′–12′*. The height of the miter line *15–9′* is made to equal *6′–9′*, or the semi-diameter of the large pipe *R*, and this distance is taken as a radius for describing the quarter circle which represents a true half section on the miter line *15–9′*, shown at *G*.

Draw the elevation and place the profiles and sections in position as shown. Space the profile *D* into a number of equal divisions, and from these points draw vertical lines, which intersect the miter line from *1′* to *5′*. At right angles to branch *A*, draw the stretchout line *5–5*, upon which place the girth of profile *D*, as shown from *5* to *1* to *5*. Draw the measuring lines which intersect by horizontal lines draw from similarly numbered points on the miter line *1′–5′*. A line traced thru these points will give the miter cut of the full pattern for branch *A*, shown at *J*. The divisions on the miter cut in pattern *J* are used in developing the upper edge line of the pattern for branch *B* in the elevation. The half profile *R* of the large pipe is now divided into equal parts, as shown from *6′* to *12′*, and from these divisions vertical lines are drawn to the base line in elevation. In corresponding manner, space the half section *G*, as shown from *13* to *15*, and draw horizontal lines which intersect the

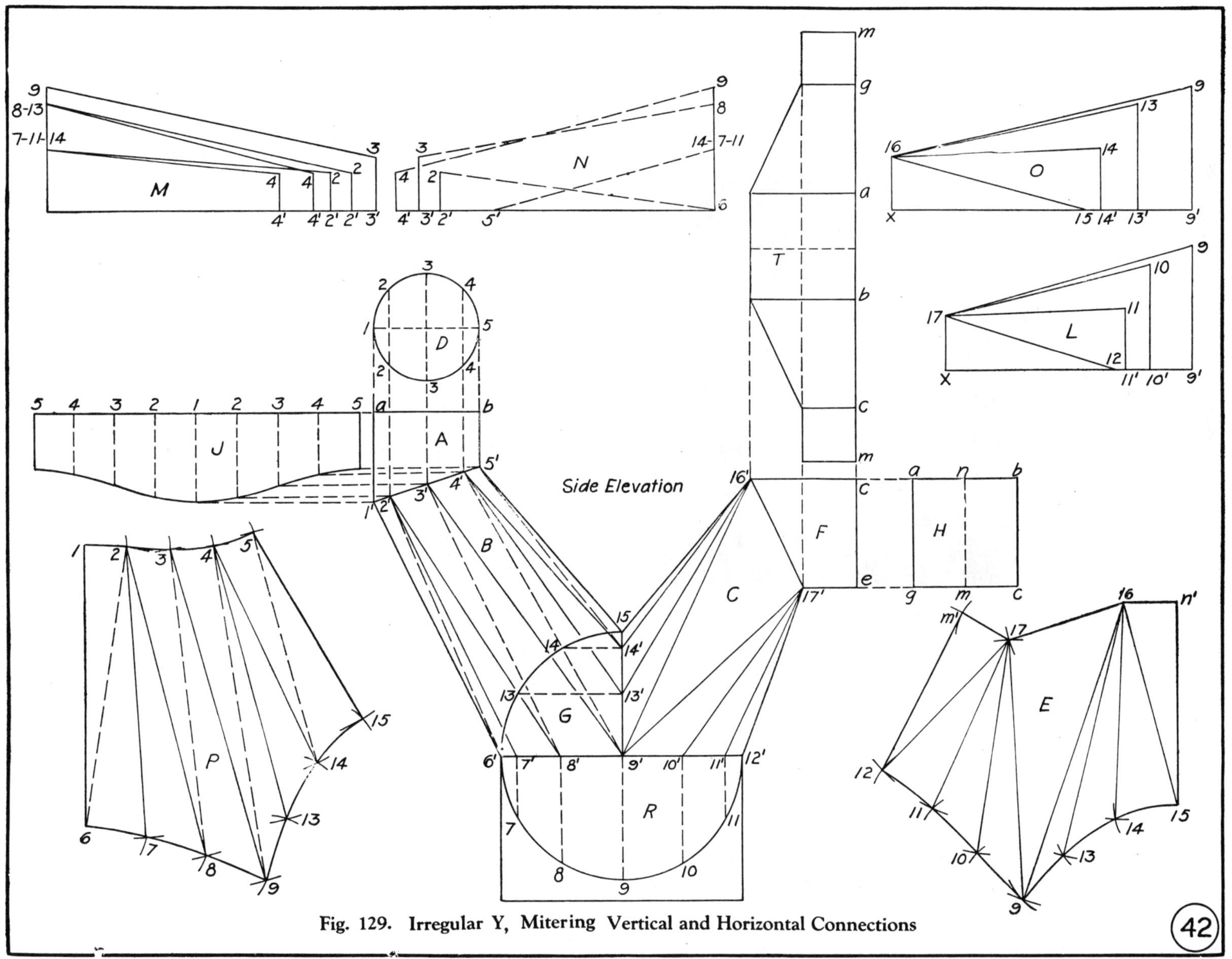

Fig. 129. Irregular Y, Mitering Vertical and Horizontal Connections

miter line *15–9′*, as shown. Now, connect the points on the upper and lower bases of the transition piece *B* by means of solid and dotted lines. The true lengths of the solid lines are shown in diagram *M*, and those for the dotted lines in diagram *N*, and are found by the method given in preceding problems.

The half pattern for branch *B* is shown at *P*. The true lengths of the solid and dotted lines are obtained from diagrams *M* and *N*, respectively. The spaces from *6* to *9* to *15* on the lower edge of the pattern are obtained from divisions of corresponding numbers in sections *G* and *R*, respectively. The divisions from *1* to *5* on the upper edge of pattern *P* are obtained from the divisions along the miter cut in pattern *J*. The same method is employed for developing the patterns for transition piece *C* and horizontal branch *F*.

The half pattern for *C* is shown at *E*; the divisions *12* to *9* and *9* to *15* on the lower edge of the pattern are obtained from similarly numbered spaces on half section *G* and half section *R* of the large pipe. The true lengths of the lines in pattern *E* are shown in diagrams *O* and *L*; the vertical lines *17–X* and *16–X* are equal in length to *mg* or *na* in profile *H*. The full pattern for branch

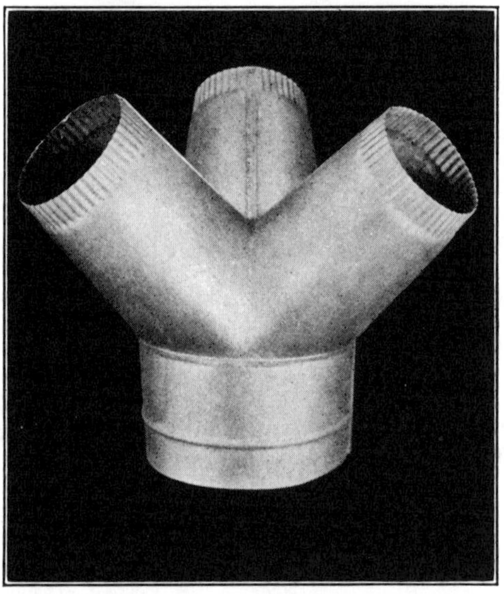

Problem 93. Three-Branch Fitting.

F is shown at *T*, and is developed by the parallel method in the usual manner.

Problem 93. Three-Branch Fitting

Figure 130 shows the method employed when developing patterns for fittings which contain three or more branches, and is also applicable, whether the branches are inclined at the same angle, or whether the openings in the branches are of the same shape or not. In this problem we have three branches, all inclined at the same angle and whose openings at the upper end are of the same diameter, so that the pattern for one branch is all that is required. The three branches are to be carried off from a 5-1/4-inch pipe in such manner that the angle made between the center line of each branch and the axis of the main pipe is one of 133°. The open ends of the branches are to be equal, each being 3 inches in diameter.

In laying out the drawing, it is only necessary that one branch be drawn at right angles to its center line *mn* in plan, as shown by *1–4′–7′–8–14* in the partial elevation. First, describe the circle *4–g–m* in plan, whose diameter shall be equal to the large pipe. The outline of the circle is next divided into three equal parts, thus locating the points *4*, *g* and *m*, and from these points draw the miter lines *m–1′*, *g–1′* and *4–1′*. From the center of the circle in the plan, draw the vertical center line *1′–1*, making *0–1* equal to the semi-diameter of the large pipe. Next, from *O* on this line, draw a line at an angle of 45°, to serve as the center line of the branch pipe. Locate point *11′* on this line, 6 inches from *O*, and erect the perpendicular *14–8*.

The half profile, shown at *B*, is constructed, and the outline of the semi-circle divided into a number of equal spaces. The points thus located are then projected to the line *14–8*, as shown. Next, thru *O* draw the line *4–7′* equal in length to the diameter

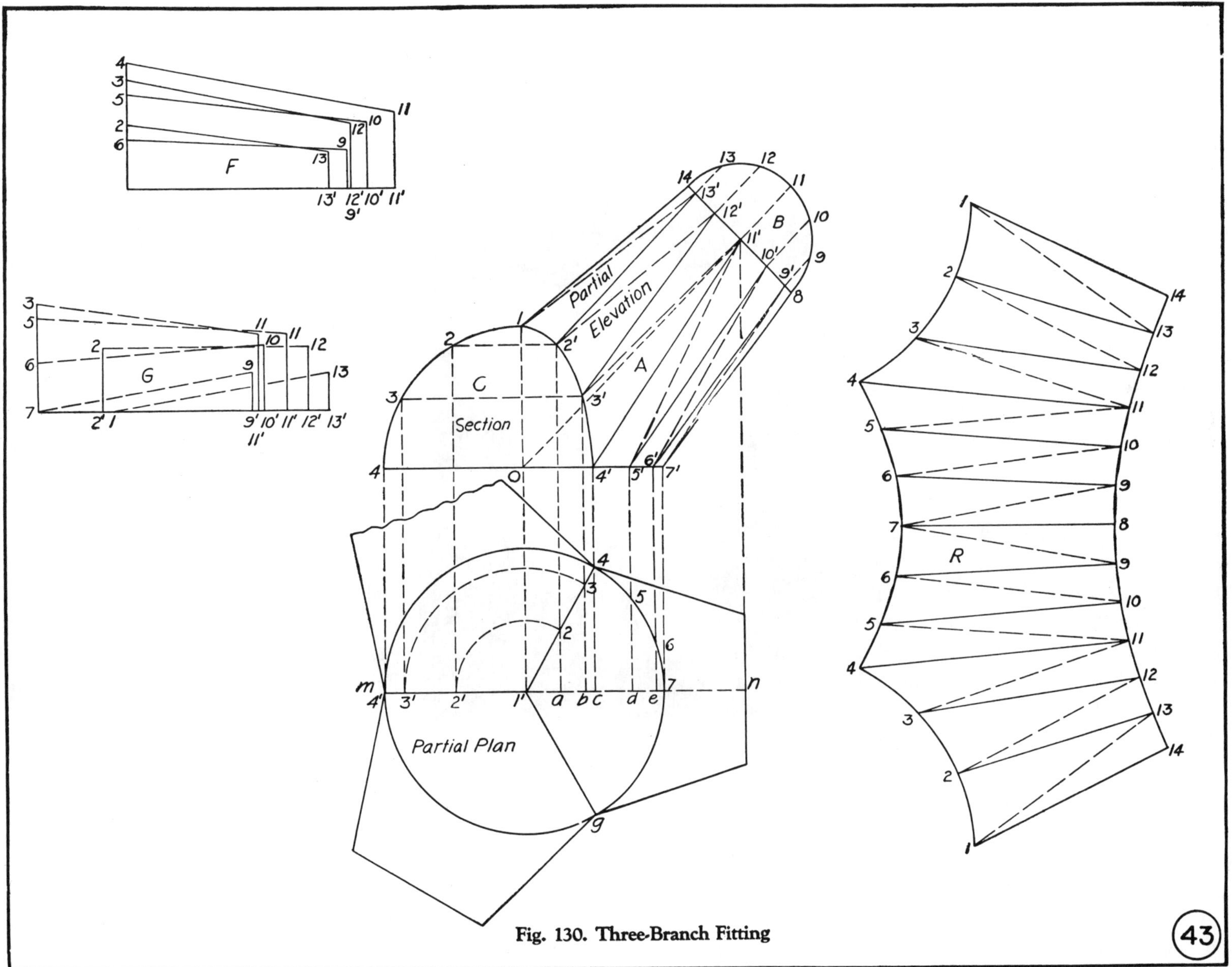

Fig. 130. Three-Branch Fitting

43

of the large pipe.   With *O* as center, and *O–4* as radius, describe the arc *4–1*.   Then *1–O–4* will represent the true section on *4′–1′* in plan, as shown at *C*.   Divide this section into equal parts, and from these divisions draw vertical lines to the miter line *1′–4′* in plan, intersecting this line at *2′* and *3′*, as shown.   With *1′* as center and radii equal to *2′* and *3′*, describe arcs which intersect the miter line *1′–4*, as shown by *2* and *3*.   From these points draw vertical lines, which are intersected by horizontal lines drawn from similarly numbered points in section *C*.   A curved line traced thru these points of intersection is then the foreshortened miter line, as shown by *1–2′–3′–4′*, in the elevation, and by the miter line *1′–4* in plan.

Now, divide the distance from *4* to *7* on the arc of the large circle in the plan into equal parts, as shown by *4, 5, 6* and *7*, and from these points draw vertical lines intersecting the base line of the fitting in the elevation at *4′, 5′, 6′* and *7′* Next, draw solid and dotted lines in branch *A* and obtain their true lengths, as shown in diagrams *F* and *G*.

To find the true lengths of the dotted line *2′–12′*, place this distance as shown from *2′* to *12′* in diagram *G*, from which draw the vertical lines *2′–2* and *12′–12*, making *2′–2* equal to the distance measured from the point *a* on the line *mn* in plan to the point *2* on the miter line *4–1′*, and the line *12′–12* equal to *12′–12* in half profile *B* in elevation.   The distance from *2* to *12* in diagram *G* is then the true length of the dotted line *2′–12′* in the elevation.

The true lengths of all the solid and dotted lines are obtained in similar manner, and the pattern developed by the method described in Problem 87.

The full pattern for branch *A* is shown in *R*.   The divisions from *8* to *14* on the upper edge of the pattern are obtained from the half profile in *B*.   The divisions from *.7* to *4* are equal to the divisions from *7* to *4* on the outline of the circle, which represents the large pipe in plan.   The divisions from *4* to *1* in pattern *R* are taken from *4* to *1* in section *C*.

Problem 94.   Three-Way Branch

In Figure 131 is shown the elevation patterns for a three-way branch fitting, and round to round.   The branches *A, B* and *C* are carried off from a 5-1/4-inch round pipe in such manner that the centers of the round openings at the upper end are in the same vertical plane, thus making both sides of the branch symmetrical.   First, draw the center line *18–11*.   From point *11′* on this line draw the center line of the branch *B* at an angle of 60°, making it 8 inches long, as shown by *11′–4′*.   At right angles to the center line of branches *A* and *B*, draw the lines *1–7* and *a–15*, making them 3 inches long, or equal to the diameter of the outlet of each branch.

Next, draw the base line *m–8* equal to the diameter of the main pipe, and place the semi-profiles in their proper position, as shown by the semi-circles *G, H* and *F*.   The next step is to

Problem 94.   Three-Way Branch.

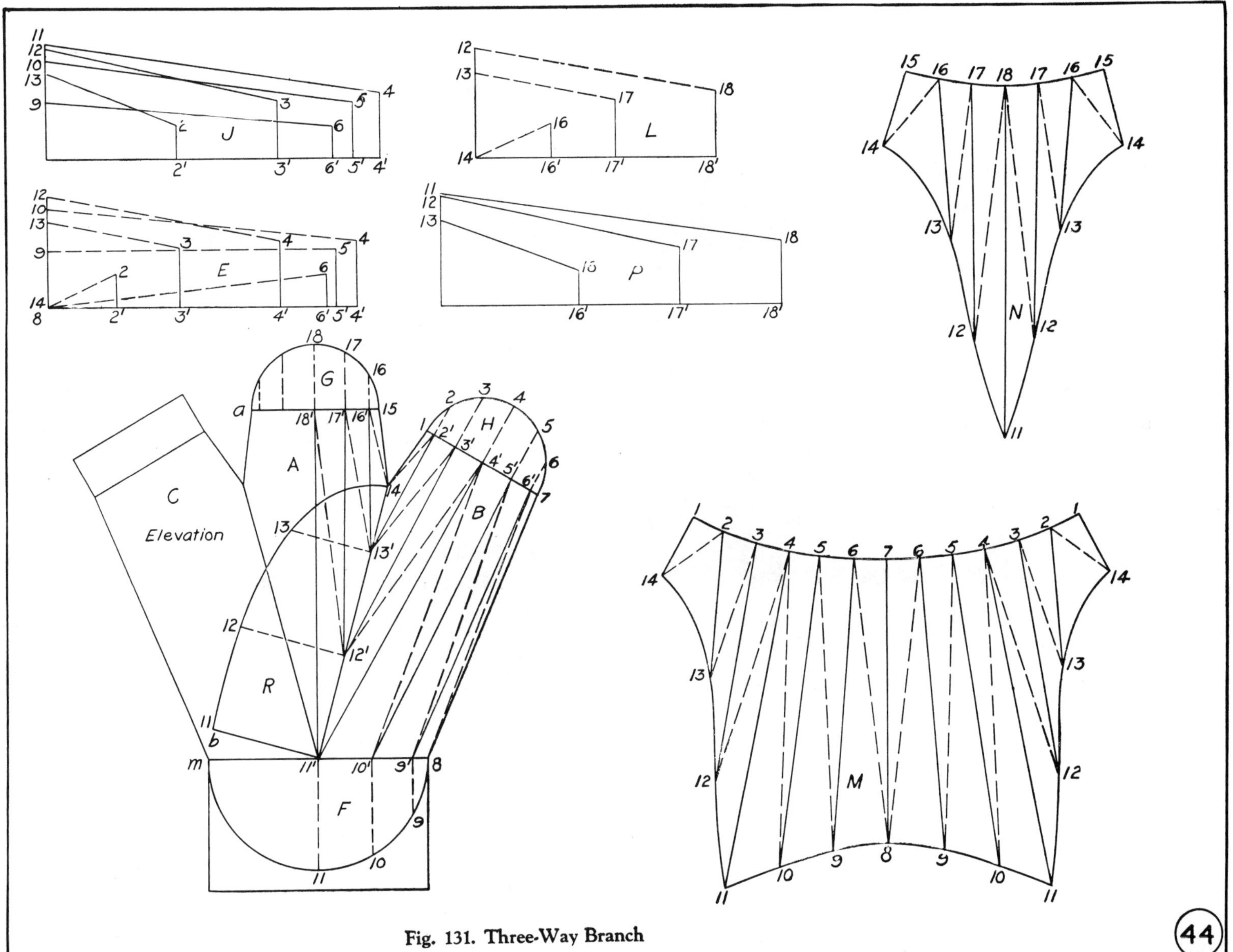

Fig. 131. Three-Way Branch

obtain a true section on the miter line *14–11′*, which is constructed in the following manner:

From point *11′* at the lower end of the miter line and at right angles to *11′–14*, draw the line *11′–b*, making it equal to the semi-diameter of the large pipe, as shown by *11′–8* in profile *F*. Connect points *14* and *b* by means of an irregular curve, which will represent a true half section on the miter line *14–11′*, shown in *R*. As the side branches *B* and *C* are alike, only one pattern will be required; also a separate pattern for the center branch *A*, both of which will

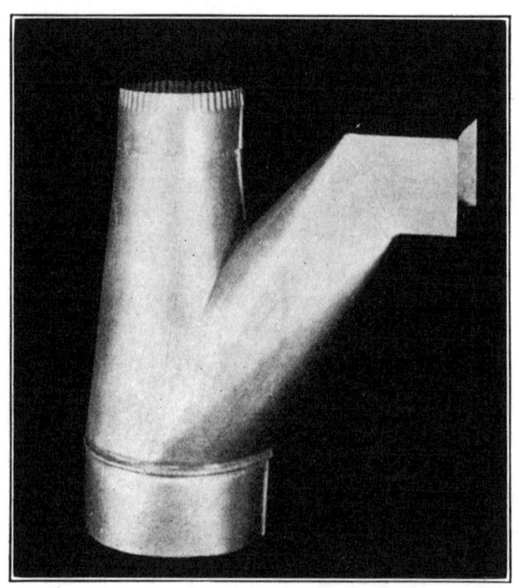

Problem 95. Two-Way Transitional Branch, Horizontal and Vertical Connections.

be developed by triangulation. To develop the pattern for branch *B*, first divide the quarter section *F* and half section *R* into three equal parts, as shown by the figures *8* to *11* and *11* to *14*; then divide half profile *H* into six equal parts, as shown from *1* to *7*. From these points, at right angles to the various base lines, draw lines intersecting the base lines, as shown by similar numbers.

Draw solid and dotted lines in branch *B* and find their true lengths by constructing the diagram, shown at *E* and *J*, by the method explained in previous problems. The full pattern for branch *B* is shown in *M*. The divisions from *7* to *1* on the upper edge of the pattern are obtained from the half profile in *H*. The divisions from *8* to *11* are equal to the divisions from *8* to *11* on quarter section *F*, and the divisions *11* to *14* are equal to the spaces on half section *R*. As the four quarters of center branch *A* are alike, divide one-half of semi-profile *G* into three equal parts, and from these divisions draw vertical lines to the center line *a–15*, as shown by *17′* and *16′*. From these intersections, draw solid and dotted lines to points *12′* and *13′* on miter lines *14–11′*. The true lengths of these lines are shown in diagrams *P* and *L*.

A one-half pattern for branch *A* is shown in *N*. The divisions from *11* to *14* are equal to the divisions from *11* to *14* on half section *R*. The divisions on the upper edge of pattern *N* are equal to the divisions from *18* to *15* in semi-profile *G*. The patterns *M* and *N* are developed in precisely the same manner as described in connection with Figure 124.

Problem 95. Two-Way Transitional Branch Mitering with Horizontal and Vertical Connections

Figure 132 presents the shortened method employed in developing the patterns for a two-way transition branch. The main pipe, shown by *e–g–15–1* in the elevation, is round in form. The opening in branch *G* is rectangular in shape and is mitered to a horizontal pipe. Branch *F* is tapering in form, the upper end being mitered to the round vertical pipe, shown by *a–b–14–8* in the elevation.

First, draw the elevation of the tapering branch *F*, and describe the semi-circles *D* and *H*, which represent the half profiles of the upper and lower bases. Divide profiles *D* and *H* into the same number of equal divisions, and project these points to the upper and lower base lines in the usual manner. Connect these

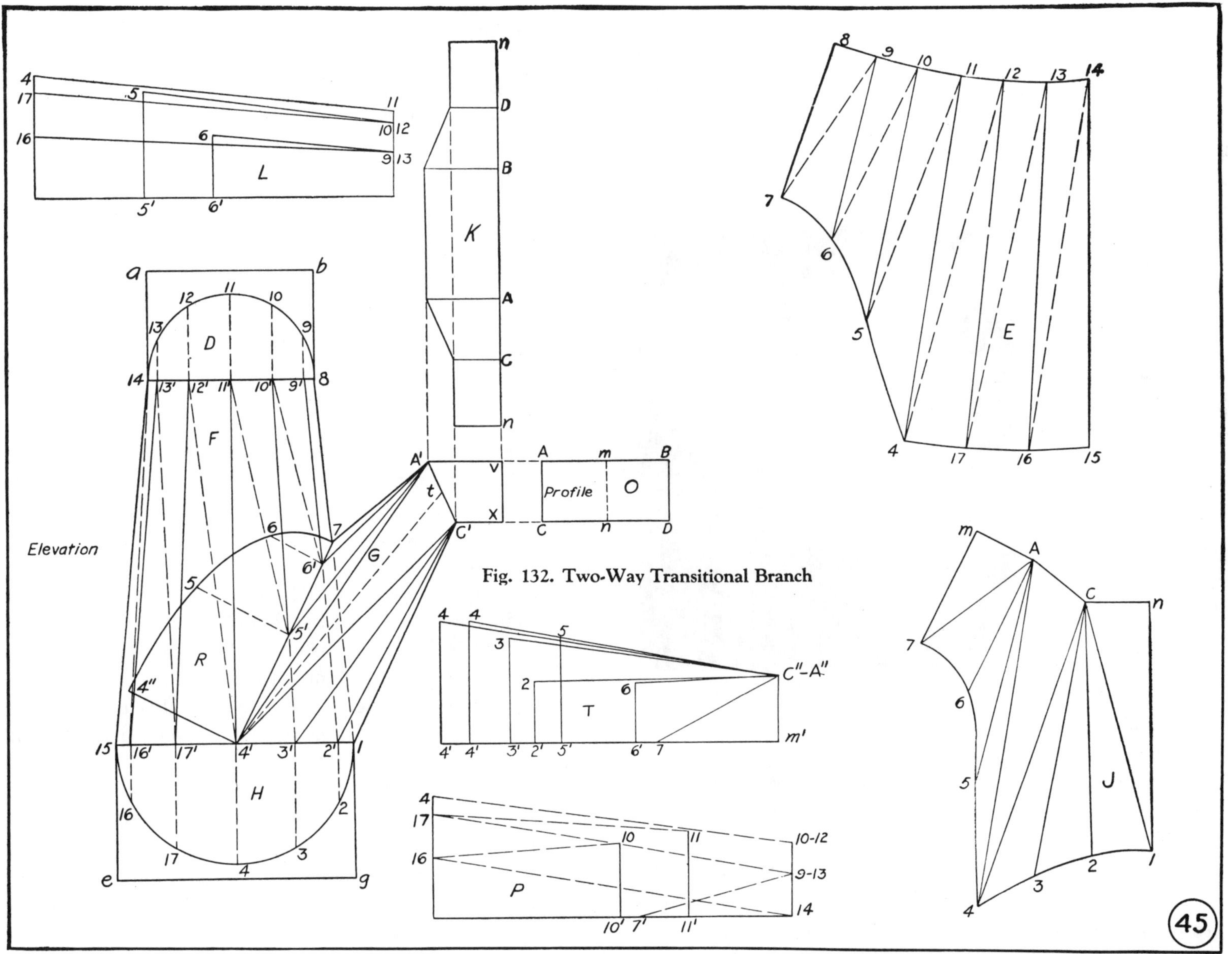

Fig. 132. Two-Way Transitional Branch

points by solid and dotted lines, as shown by *10'–3'*, *9'–2'*, etc., in the elevation. Next, from point *4'* on the lower base line of branch *F*, draw the center line *4'–t* of branch *G* at an angle of 50° to the base line, and draw the outline of branch *G*, as shown by *4'–7–V–X–1*. Draw the profile of the rectangular pipe, shown by *A*, *B*, *C* and *D*; also the half section on the joint or miter line *7–4'*, shown at *R*.

From points *5'* and *6'* on the miter line, and *3'* and *2'* on the lower base line, draw lines to the corners *A'* and *C'* in branch *G*, as shown. The true lengths of these lines are shown in diagram *T*, the distance *m'c''* being equal to one-half the width of the rectangular pipe, shown by *mA* and *nC* in the profile.

The true lengths of the solid and dotted lines in the tapering branch *F* are shown in diagrams *P* and *L*, and a one-half pattern for same is shown in *E*. The spaces from *14* to *8* on the upper edge of the pattern *E* are equal to the spaces from *14* to *8* in half profile *D*. The spaces from *15* to *4* on the lower edge are equal to the spaces from *15* to *4* on the half profile of the main pipe, shown in *H*. The spaces from *4* to *7* are equal to the spaces from *4''* to *7* on half section *R*.

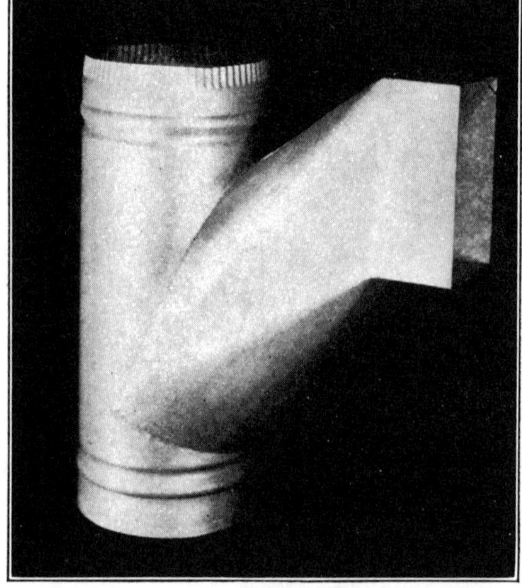

Problem 96. Irregular T-Joint, Rectangular to Round.

A one-half pattern for transitional branch *G* is shown in *J*; the divisions from *M* to *n* on the upper edge are taken from profile *O*. The divisions from *1* to *4* and *4* to *7* are equal to correspondingly numbered divisions on half profile *H* and half section *R*. The full pattern for the horizontal rectangular branch is shown at *K*, and is developed by the parallel-line method.

Problem 96. Irregular T-Joint, Rectangular to Round

Figure 133 shows the simplified method employed in developing the patterns for an irregular T-joint, changing from a round to a rectangular outlet. A plain view is here shown, but is not required for the development of the pattern, as both valves are symmetrical.

First, draw the side elevation in which *BDFH* represents the round pipe. From point *4'* on the center line, draw the line *4'–C*, which represents the angle of the T on its center line, and construct the elevation of the transition piece, as shown by *4'–1''–G'–A'–1'*.

The section or profile of the rectangular pipe on the line *G'–A'* is indicated by *SFRA* in profile *E*, thru which center line *ab* is drawn. Next, draw half profile *H* of the round pipe. As both quarter circles are symmetrical, it is only necessary to divide the quarter circle into an equal number of spaces, as shown by the figures *1* to *4*, and from these points draw vertical lines, which intersect the miter lines in the elevation at points marked *1'* to *4'* and *4'* to *1''*. From the intersections on the miter line *4'* to *1''*, draw lines to *G'*, and from the divisions on the miter line *4'* to *1'* draw lines to *A'*. These lines then represent the bases of sections having altitudes equal to the heights in the semi-profiles *E* and *H*, and their true lengths are found by constructing the diagrams of triangles, shown at *R* and *L*. For ex-

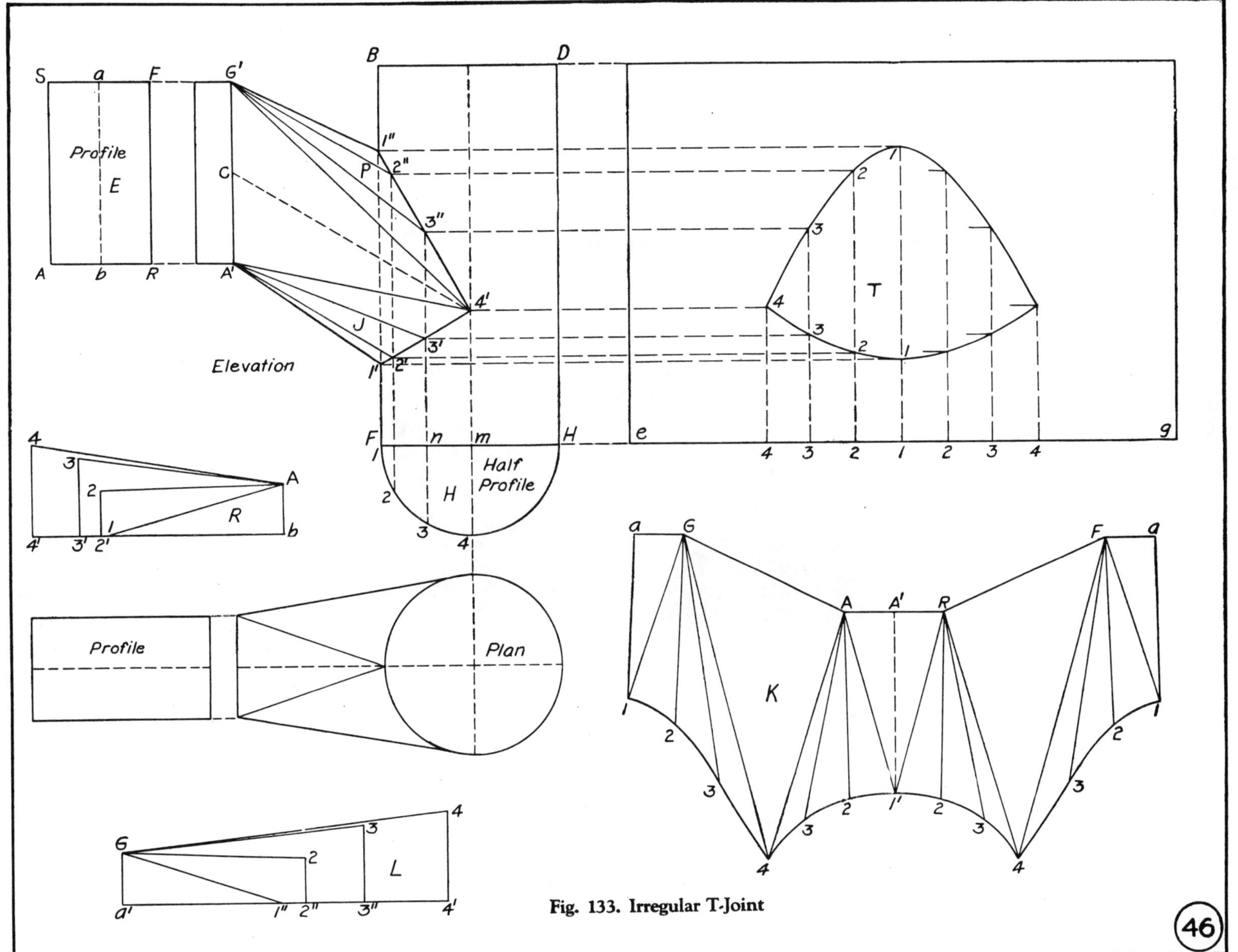

Fig. 133. Irregular T-Joint

ample: To find the true length of the line *4′–G′* in the elevation, take this distance and place it on any line as *a–4′* in diagram *L*; draw the perpendicular *a′–G* and *4′–4* equal in height to *S–a* in profile *E*, and *4–m* in profile *H*. A line drawn from *G* to *4* in the diagram will be the true length of the line *G′–4′* in the elevation. Again, for an example: To find the true length of the line *A′–3′* in the elevation, take this distance and place it on the base line in diagram *R*, as shown from *b* to *3′*. Draw the vertical lines *b–A* and *3′–3*, equal, respectively, to *Ab* in profile *E*, and *3–n* in semi-profile *H*. A line drawn from *3* to *A* will be the true length of this line. In this manner, all of the true lengths are obtained. Diagram *R* shows the true lengths of the lines in *J*, and diagram *L* the true lengths of the lines in *P*.

The pattern for the opening in the vertical pipe is shown in *T*, from which the stretchout of the lower edge of the pattern for the transition piece can be obtained. To obtain the pattern for the opening in the vertical pipe, take the stretchout of the semi-profile *H* and place it on the line *eg*, as shown by the numbers *4* to *1* to *4*. From these divisions, draw vertical lines, which are intersected by horizontal lines drawn from similarly numbered intersections on the miter line in the elevation. A line traced thru these points of intersection will give the pattern for the opening to be cut in the round pipe to receive the transition piece. The full pattern for the transition piece is shown in *K*.

The distance *A′–1′* and *a–1* are equal to *G′–1″* and *A′–1′* in the elevation, which are shown in their true lengths. The distances *aG*, *GA*, *AR*, etc., on the upper edge of pattern *K* are obtained from the profile of the rectangular pipe, shown at *E*. The true lengths from the diagrams *R* and *L*, and the divisions on the lower edge of the pattern are obtained from similarly numbered divisions in the pattern for the opening in the vertical pipe, as shown in *T*.

Problem 97. Two-Way Elbow Fitting

The sheet-metal worker engaged in the construction of fittings and transition pieces in pipes used for exhaust and heating systems has frequent use for a two-way elbow fitting, shown in Figure 134. The two branches are made in the form of five-piece elbows which are connected to the main pipe by a transition piece, shown by *m–A–B–1* in the elevation. The arrangement of the fitting is such that the elbow sections *C* and *C′* intersect each other for a certain distance on the vertical center line, as shown by the line *ag*.

First, draw the elevation of the elbows in correct position, the distance *m–1* being made equal to the diameter of the main pipe. Next, draw the lateral outlines of the transition piece which are represented by the vertical lines *mA* and *B–1* in the elevation. The line *AB* is then drawn, completing the elevation of the transition piece, to which is added a straight collar 2 inches wide,

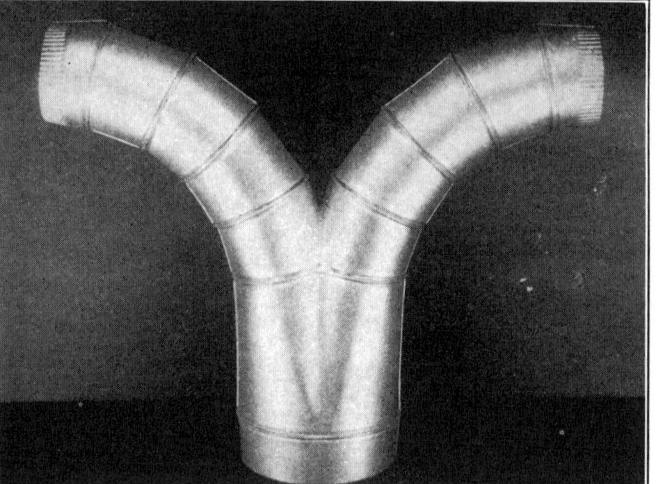

Problem 97. Two-Way Elbow Fitting.

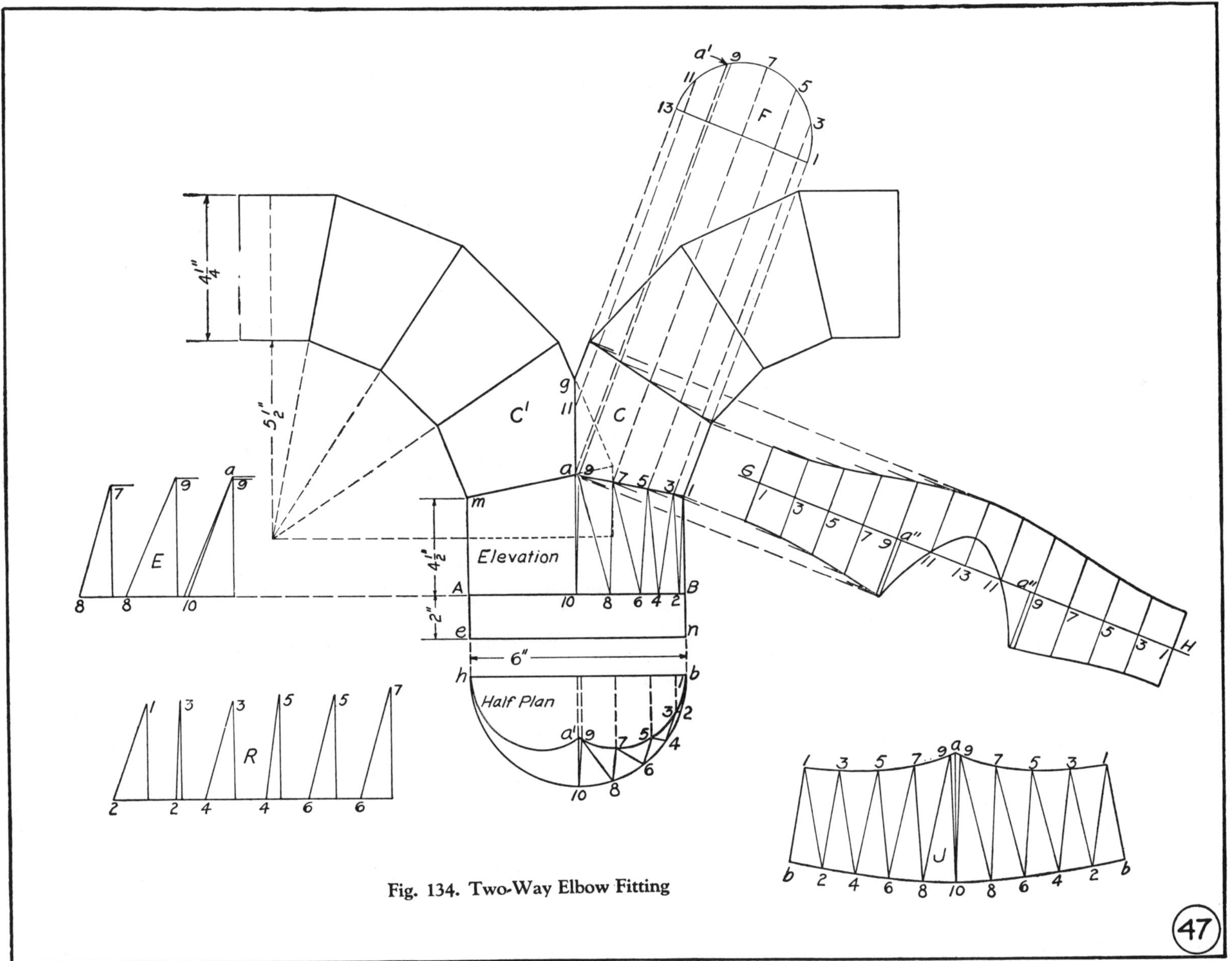

Fig. 134. Two-Way Elbow Fitting

shown by *A–e–n–B*. As both elbows are of equal diameter, the pattern for section *C* is the only pattern required; the patterns for the entire elbow may be derived therefrom. The center line of section *C* is drawn and extended to the upper portion of the drawing; next, on this line construct the half profile of the elbow, shown by the semi-circle at *F*. Divide the semi-circle into equal spaces, and from these points, draw lines parallel to the center line across the elevation of the elbow until they intersect the miter lines *ga* and *a–1* in section *C*.

The pattern for section *C* may now be developed by setting off the stretchout on the line *G H* in the usual manner, and the point *a* on the miter line is located at the point *a'* on the half profile at *F*, and also at *a''* on the stretchout line *G H* in the pattern. The upper line of this pattern will serve for the remaining sections of the elbow. The pattern for section *C* of the elbow should always be developed first, because the distances between spaces on the lower miter cut of the pattern will be required in obtaining the length along the upper edge of the transition piece in its development.

The transition piece is shown in the elevation by *A–B–1–a–n*. From the half plan, shown below the elevation, it will be seen that the lower base of this piece has for its outline a circle, shown by *h–10–b*, and that its upper base on the lines *ma* and *a–1* are parts of two circles that represent the lower ends of the elbow sections *C* and *C'*, whose position upon the plan is found in the following manner:

From each of the points on the miter line *a–1* in the elevation, draw vertical lines to the plan, crossing the center line *hb*. Measuring from line *hb*, each line is made equal to the distances from the center line *1–13* of half profile *F* to the points *1, 3, 5, 7,* etc., on the semi-circle. A line traced thru these points will show the outline of the upper base in plan. Next, the outline of

the lower base in the plan is divided into the same number of equal parts.

The triangles may now be indicated in the plan by drawing lines between successive points on the upper and lower bases, as *2–3, 3–4, 4–5, 5–6*, etc., and afterwards projecting these lines to the elevation.

The triangles on the surface of the transition piece in the elevation are necessary only to assist in seeing some of the elements which appear somewhat confused, as seen upon the plan.

The true lengths are found in the customary manner, as shown by the diagram of triangles in *R* and *E*.

A one-half pattern of the transition piece is shown at *J*, and the method of laying out the pattern is similar to preceding developments. It is necessary merely to note that the distances *a–9, 9–7, 7–5, 5–3* and *3–1* on the upper edge of the pattern are taken from similarly numbered divisions on the miter cut of the elbow pattern, the distances being there shown in their true lengths.

The divisions from *10* to *b* on the lower edge of pattern *J* are equal to the spaces from *10* to *b* in the half plan.

Problem 98. Ship Ventilator with Round Mouth and Base

In Figure 135 is shown the simplified method of developing the patterns for a round tapering elbow. An elbow of this form is generally known as a ship ventilator, and the principles shown in this problem are applicable to any form or shape, no matter what the respective profiles may be at the base or top.

Ship ventilators of this type are used on sailing vessels and steamships of all kinds, and are made in a great variety of forms and in proportions to suit the various uses and accompanying conditions. No rule can be given to suit all conditions, but the

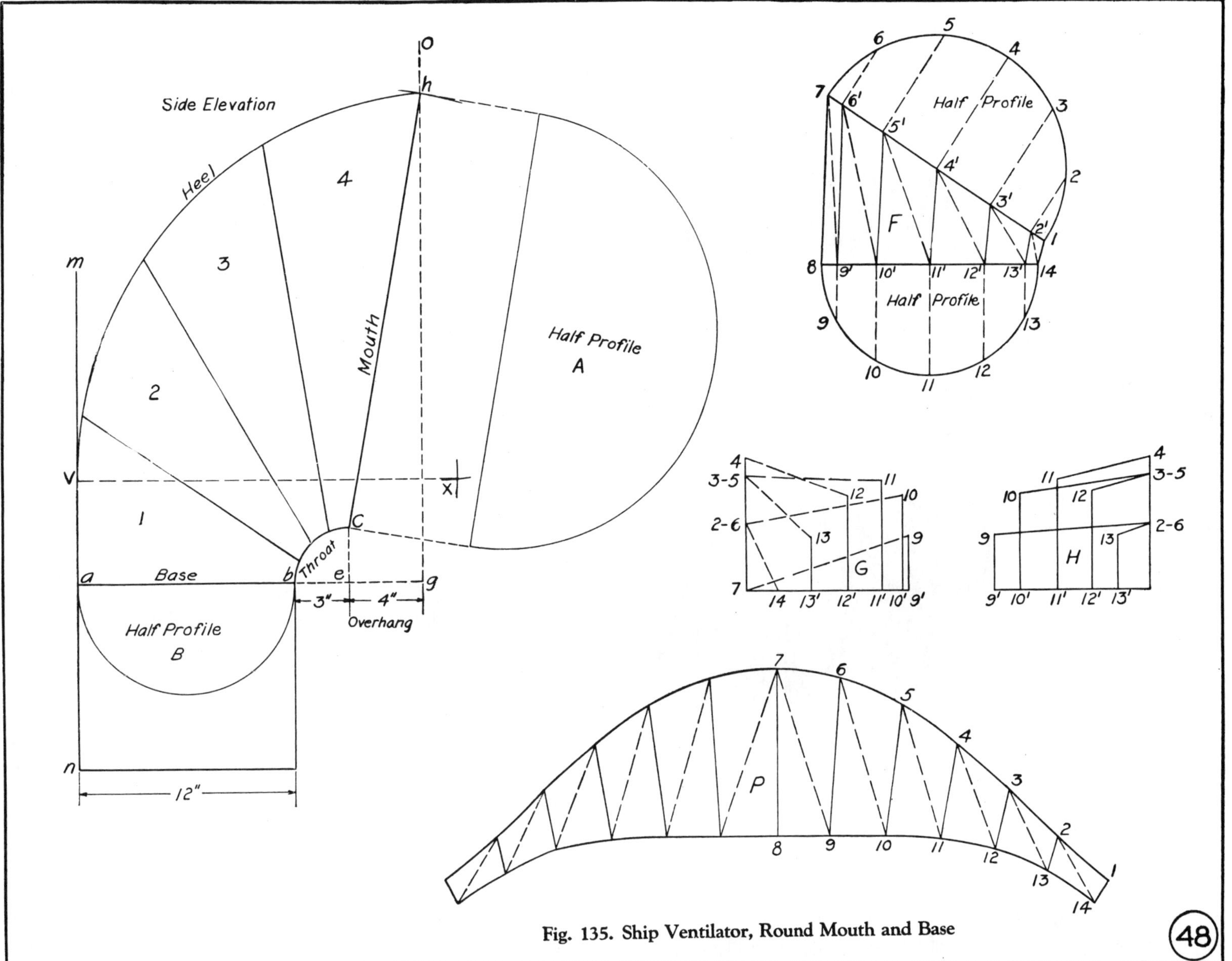

Fig. 135. Ship Ventilator, Round Mouth and Base

48

general proportions given in this problem may be taken as standard proportions for designing any size ship's ventilator of this type.

Figure 135 shows a round ventilator made up in four sections; the half profiles of the upper and lower openings are shown in the elevation by the semicircles *A* and *B*. The size of the base or lower opening constitutes the basis for determining the outline of the throat and heel and the proportions of the ventilator. In this case, assuming the diameter of the base to be 12 inches, draw the base line 12 inches long, establishing the points *a* and *b*. From point *a* erect a line perpendicular to the base line, extending it indefinitely, as shown by the line *mn*. The radius of the throat is next determined by taking one-fourth the diameter of the base, or 3 inches, which is marked off on the base line extended to the right from point *b* to point *e*. Using *e* as center and *eb* as radius, describe the quarter circle, or throat, line *bc*, and intersect it by the perpendicular line erected from *e* at *C*. As the diameter of the mouth of the ventilator should be twice

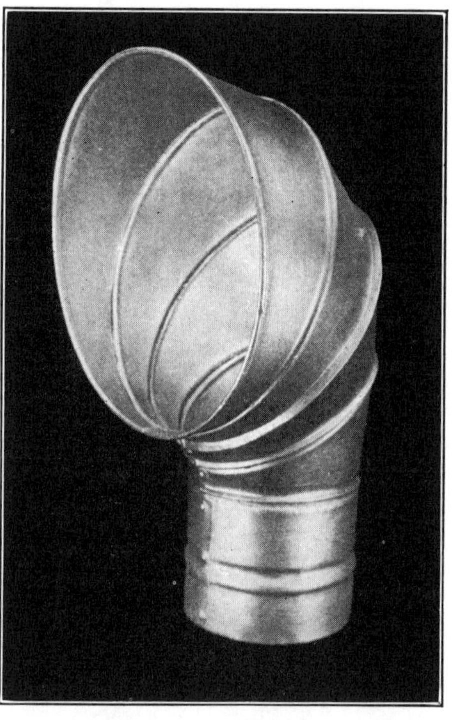

Problem 98.   Ship Ventilator, Round Mouth and Base.

the diameter of the base, and the overhang one-third the diameter of the base, the elevation is completed in the following manner: From point *e* on the base line, set off *eg* to equal one-third the diameter of the base, or 4 inches.   From *g* at right angles to *eg*, draw the vertical line *go*.   Now, with *c* as center and twice the diameter of the base, or 24 inches, as radius, describe a short arc, intersecting the vertical line *go* at *h*.   Draw the inclined line *hc*, which represents the mouth of the ventilator.

The next step is to draw the heel of the elbow.   Using a radius equal to one and three-fourths times the diameter of base, and with one leg of the dividers on point *h* and the line *mn*, describe short arcs intersecting at *x*.   With *x* as center and *xh* as radius, draw the arch *hv* tangent with the vertical line *mn*.   This method, in the form of a rule, shows: (1) *Throat radius* equals one-fourth diameter of base.   (2) *Heel radius* equals one and three-fourths times diameter of base.   (3) *Mouth diameter* equals twice diameter of base.   (4) *Overhang* equals one-third diameter of base.

The throat and heel are now divided into four equal parts, and these points connected by lines, which represent the miter lines of each section which is developed by triangulation.   To obtain the pattern for section Number 1 in the side elevation, take a tracing of this section and place it at the right of the elevation, as shown by *7–1, 14–8* in *F*.

Bisect *7–1* and *8–14* and obtain the points *4'* and *11'*, respectively.   With these points as centers, describe a semi-circle on each side, which will represent a half profile of each end of the section.

Next, divide these half profiles into the same number of parts, as shown by the figures *1* to *7* and *8* to *14*.   At right angles to *1–7* from the various points on the half profile, draw lines to intersect the line *1–7* at *2', 3', 4', 5'* and *6'*.   Also, at right angles

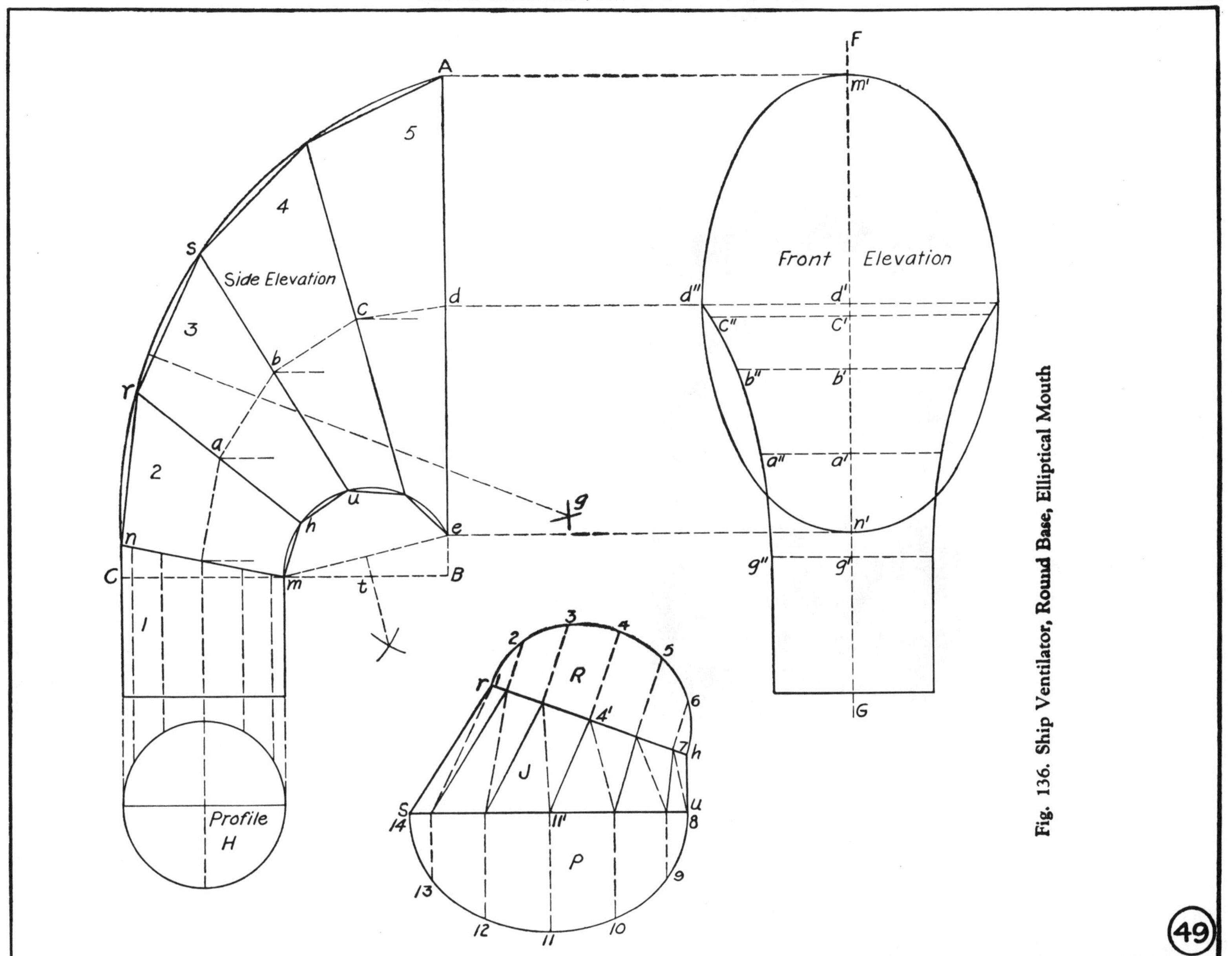

Fig. 136. Ship Ventilator, Round Base, Elliptical Mouth

49

to the base line *8–14* from the divisions *9, 10, 11, 12*, etc., on the half profile, draw lines to intersect the line *8′–14* at *9′, 10′, 11′, 12′* and *13′*. Draw solid lines from *6′* to *9′*, *5′* to *10′*, *4′* to *11′*, *3′* to *12′* and *2′* to *13′*; also the dotted lines *7* to *9′*, *6* to *10′*, *5* to *11′*, etc. The true lengths of the solid and dotted lines are obtained in the usual manner by constructing the diagram of triangles, shown in *G* and *H*. The completion of the pattern for section Number 1 is shown in *P*, and since the process is precisely similar to that described in preceding problems that have been developed by the simplified method of triangulation, no further mention of the principles involved is necessary.

A construction similar to that shown in *F* is required for each section of the ventilator. After the patterns for each section have been developed, as shown in *P*, double edges must be allowed to the lower part of the patterns for sections 4, 3 and 2, and single edges to the upper part of the patterns for sections 1, 2 and 3. In a similar manner, the patterns for a ship ventilator or tapering elbow, composed of any number of sections, may be constructed.

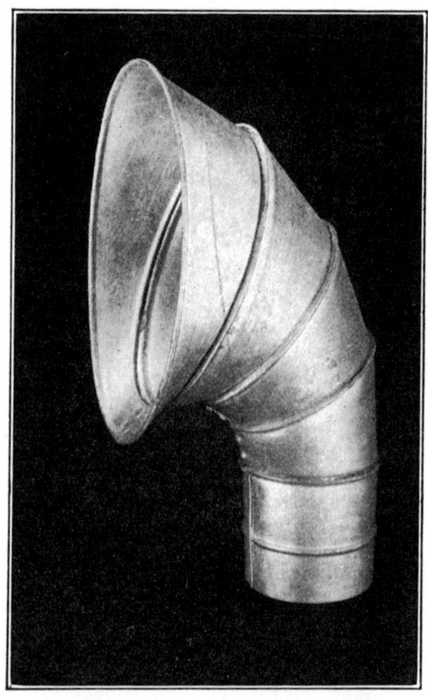

Problem 99. Ship Ventilator, Round Base and Elliptical Mouth.

Problem 99. Ship Ventilator with Round Base and Elliptical Mouth

Figure 136 illustrates how the patterns are developed for a ship ventilator or any other form of tapering elbow, making a transition from one profile to another.

The profile of the base is in the form of a circle, while the mouth is defined in the front elevation by an ellipse.

First, draw the side elevation, as this view shows the outline and principle proportions of the ventilator. In this problem it may be assumed that the diameter of the base is 6 inches, and the overhang will equal the diameter of the base. Construct the right angle *ABC*. From the point *B* on the horizontal line *BC*, a distance of 6 inches is measured and the point *m* is located, as shown. The diameter of the lower opening is then measured off on the same line from the point *m* to point *C*. From the points *m* and *C*, vertical lines are drawn and the circle is then described, which represents the profile of the lower end of the ventilator. Next, from point *m*, the inner side of the elbow, draw the miter line for a five-piece elbow in the usual manner, as shown by the line *mn*.

The lower extremity of the elliptical mouth of the ventilator is located by making the distance *Be* equal to one-fourth the diameter of the base. The upper point is located at *A* and the distance *eA* is equal to two and three-fourths times the diameter of base, or 16-1/2 inches. With this distance as radius and with *A* and *n* as centers, describe short arcs intersecting at *g*. With the point *g* as center, and with *gA* as radius, describe the arc *nA*, which represents the heel of the ventilator. Now, with *t* as center and *te* as radius, describe the arc *em*, which shows the lower outline of the ventilator. The arcs *nA* and *em* are next spaced into four equal parts, and chords are drawn between the points located on the outline of each arc. Complete the elevation by drawing the miter or joint lines of each section, and bisect each line, locating the centers *a, b, c* and *d*.

The next step is to draw the front elevation, altho in actual shop practice it is but necessary to draw a half elevation. Then draw any vertical line, as *FG*. From the points *A* and *e* in the side elevation, draw horizontal lines at right angles to *FG*, intersecting *FG* at *m'* and *n'*. Take half the diameter of the round base and set it off from *g'* to *g''*.

Since it is customary to make the mouth of the ventilator of such proportions that the minor axis of the ellipse is equal in length to two-thirds that of the major axis, the ellipse shown in the front elevation may now be constructed. Take half the length of the minor axis and set it off from *d'* to *d''*. Having determined the length of the major and minor axes, the ellipse, shown in the front elevation, may now be constructed by means of the short rule described in Figure 26. Next, describe a graceful curve from *d''* to *g''*; these arcs form the lateral sides of the ventilator.

The next step consists of bisecting each of the miter lines in the side elevation, shown by *a, b, c,* and from these points at right angles to *FG*, draw horizontal lines, which intersect the line *FG*, and cross the curve *d''g''* at *a'', b''* and *c''*. The miter lines in the side elevation represent the major axes, and the horizontal distances in the front elevation the minor axis of elliptical sections to be constructed on the several miter lines in the side elevation.

The pattern for lower section Number 1 is developed by the parallel-line method; the stretchout of this pattern is obtained by dividing the outline of profile *H* into a number of equal spaces, as shown.

The patterns for sections 2, 3, 4 and 5 are developed by the simplified method of triangulation in precisely the same manner as section Number 1 of the round ventilator explained in the preceding problem. A partial development of section Number 3 in the side elevation is shown in the drawing, and a similar course is necessary for the development of each section of the ventilator. Take a tracing of section Number 3 and place it, as shown, by *r, h, s, u,* in *J*; altho in a different position from that shown in the elevation, it is merely copied same size from the latter view. Next, on the line *su* in *J*, which is the major axis, construct the semi-elliptical profile, shown in *P*, the minor axis *11'–11* being taken from the front elevation, where it is represented by the line *b'–b''*. In the semi-profile *R*, the major axis *rh* is equal to the miter line *rh* in the side elevation, and the one-half minor axis *4'–4* is equal to *a'a''* in the front elevation.

To save time, the profiles of the upper and lower bases of each section may be drawn by circular arcs, as shown in Figure 26. Divide both profiles *R* and *P* into the same number of equal spaces, as shown by the figures *1* to *7* and *8* to *14*. At right angles to the lines *rh* and *su*, from the various divisions on profiles *R* and *P*, draw lines intersecting *rh* and *su*, as shown. The surface of section *J* is now divided into triangles by means of solid and dotted lines drawn between successive points on the base lines *rh* and *su*. The true lengths of these lines are determined by constructing a diagram of triangles, and the pattern developed by the method already explained in previous problems in this chapter.

Allowance of material must be made on all patterns for seaming or riveting.

CHAPTER IX

Skylights

Skylights are known by their form as flat, double pitch, gable and hipped. The construction, shown in the following problems, may be adapted to all forms of metal skylights. When designing a skylight to suit the special requirements of the particular case in hand, the draftsman may encounter conditions that require the several parts to have profiles different from those given here, but the method of development will not differ from that given in the various problems in this chapter.

The skylight bars must be constructed in such a manner as to furnish a supporting surface for the glass, and a gutter to carry off the water of condensation. It is also important that the draftsman, when designing a skylight of any form, design the profiles of the different parts of ample size and as simple in form as consistent with required strength to allow of rapid forming into shape on the cornice brake.

Problem 100.   Flat Skylight

Figure 137 shows the longitudinal section and patterns for the most common style of flat skylights, those which are set on a curb projecting above the roof, to insure imperviousness to storms, the inclined roof having the necessary pitch. Some sheet-metal workers are under the impression that skylight developments are very difficult. This is not so, as skylight pattern-drafting entails no intricate methods of development, the patterns being obtained by the parallel-line method, and becomes a very simple matter if this method is thoroughly understood.

A sketch of the plan view is shown at F, the miter between the upper and side curbs being simply a square return miter, while the miter between the lower and side curbs is a plain butt miter. The longitudinal sectional view in Figure 137 is drawn to a scale of 4 inches to the foot, showing the section or profile of the common bar A, the lower curb B, and the upper and side curbs C. The common bar is of the universal type, and the upper and lower curbs are formed to coincide with the general dimensions of this bar, as shown.

To obtain the pattern for the side curbs of the skylight, draw the stretchout line SV, and upon this line place the stretchout of profile C, as shown by the numbers from 1 to 10. Thru these points at right angles to SV draw the usual measuring lines.

These are intersected by vertical lines drawn from similarly numbered points on the profile of side curb C. Connect the points of intersection thus obtained, which will give the pattern for the square return miter cut on the upper end of the side curb, laps being allowed, as shown by dotted lines. This miter cut without the laps is also the pattern for the miter cut of the upper curb, the cut of this pattern being the same at both ends.

The miter cut on the lower end of the pattern for the side curb is shown at G, and is simply a plain butt miter and is obtained by drawing vertical lines from the points in profile B of the lower curb to the horizontal measuring lines in pattern G, as shown. From points 8, 9 and 10 of profile B draw a vertical line which will intersect the measuring lines 8, 9 and 10 of pattern G; also from point g of profile B to line 7 of the pattern, as shown. Profile B from g to a will butt against the upright member of profile C from 7 to 6. The cut on the side curb from 7 to 6 must be a duplicate of that part of profile B from g or e to c. It is necessary then to take the distance from e to d and place it from 7 toward 6 on the stretchout line SV of the pattern, as shown at d'. From this point draw the measuring line which is intersected by vertical lines drawn from d and c of profile B, giving the required cut. The rest of the cut of the side of profile C is straight to 5; then from 5 to 4 it goes back the distance cd of profile B, and is then straight from 4 to 3.

The gutter of the side curb, shown by 3–2–1 in profile C, is cut

Problem 100.   Flat Skylight.

off at an angle, as shown at xy. This is to allow the condensation in the gutter of the side curb to drain into that of the lower curb, shown by fe in profile B. Lines projected from xy to 1–2 in the pattern will complete the miter cut for the lower end of the side curb.

A stub or short pattern of common bar A is shown at F, and is obtained in the following manner: At any convenient distance to the right and left of profiles C and B in the longitudinal section, draw the vertical measuring lines mn and ho, as shown. Next, draw the vertical line m'n' in pattern F, and upon this line place the stretchout of the common bar, or profile A, as shown by the numbers 1 to 6 to 1. Thru these points draw the usual measuring lines. Now, measuring from the line mn in the longitudinal section, take the various distances from the line to the points 5, 6, 4, 3, 2 and 1 in profile C, and place them upon similarly numbered measuring lines in pattern F, in each case measuring to the left from the stretchout line m'n', as shown. Connect these points, which will give the required miter cut upon the upper end of pattern for the common bar, laps being allowed, as shown by the dotted lines.

The miter cut on the lower end of the common bar is obtained in precisely the same manner. Measuring from the line ho in the longitudinal section, take the various distances from this line to the points on the lower curb, profile B, and place these distances on similarly numbered measuring lines to the right of the vertical line h'o' in pattern F, thus obtaining the miter cut for the lower end of common bar *A*, where it intersects the lower curb *B*.

The pattern for the miter cut on both ends of the lower curb will be the same, as shown in *H*, and is obtained by placing the stretchout of profile *B* upon the line *h″o″* and drawing the usual parallel measuring lines, as shown. Then, measuring from the line *ho* in the sectional view, take the horizontal distances from this line to the points *10, 9, 8* and *g* in profile *B*, and place them upon similarly numbered measuring lines in pattern *H*, in each case measuring from the stretchout line *h″o″*. Connect these points, and as the lower curb from *g* to *a* will be a straight cut where it will butt against the upright member *7–6* in profile *C*, draw a vertical line from point *g*, completing the pattern, as shown.

Problem 101.　Double-Pitch Skylight.

Problem 101.　Double-Pitch Skylight

Double-pitch, or gable, skylights (Figure 138) are used where a hipped skylight would be too expensive. In this form of skylight, the ridge bar is extended to reach its full length, thus doing away with the end slopes and forming what is called a gable end. The gable ends are made of metal and, sometimes, of wood. Ventilators and louvres of any suitable form are frequently placed in the gable ends. The common bar is supported at the top by the ridge bar and at the bottom on a curb, so designed as to receive the bar and to form a gutter to carry off the water that is emptied into it by the smaller gutters of the

common bars. The pitch, or slant, of the skylight should never be more than a one-half pitch, altho a one-third pitch is well adapted to this form of skylight, and is the pitch in general use for both gable and hipped skylights.

Figure 138 shows the half end elevation, profiles of the various parts, and patterns for a gable skylight having a one-third pitch. A sketch of plan view showing the position of the different members of the skylight and the curb dimensions, is shown in *C*. To construct the half end elevation, first draw the center line *6–e*, and at any convenient point on this line locate the point *g*. From *g* at right angles to the center line *6–e*, draw the line *gc* equal in length to one-half the width of the skylight, in this case 9 inches, thus locating the lower end of the glass line in common bar *B* at point *c*, as shown in lower curb *F*.

The upper end of the glass line intersects the center line *6–e* at point *5*, which determines the pitch of the skylight, and is located in the following manner: (In this case, the skylight is to have the regulation pitch of one-third.) The line *g–c* is equal to 9 inches, or one-half the width, or span. Double 9, which makes 18 inches, and divide by 3, the required pitch, which gives 6 inches. Now, measuring from point *g*, place this distance on the center line, locating point *5*, and draw the hypotenuse of the right-angle triangle *5–g–c*. Then, the line *c–5* will represent one-third pitch.

If, however, it was desired to make it one-half, or any other,

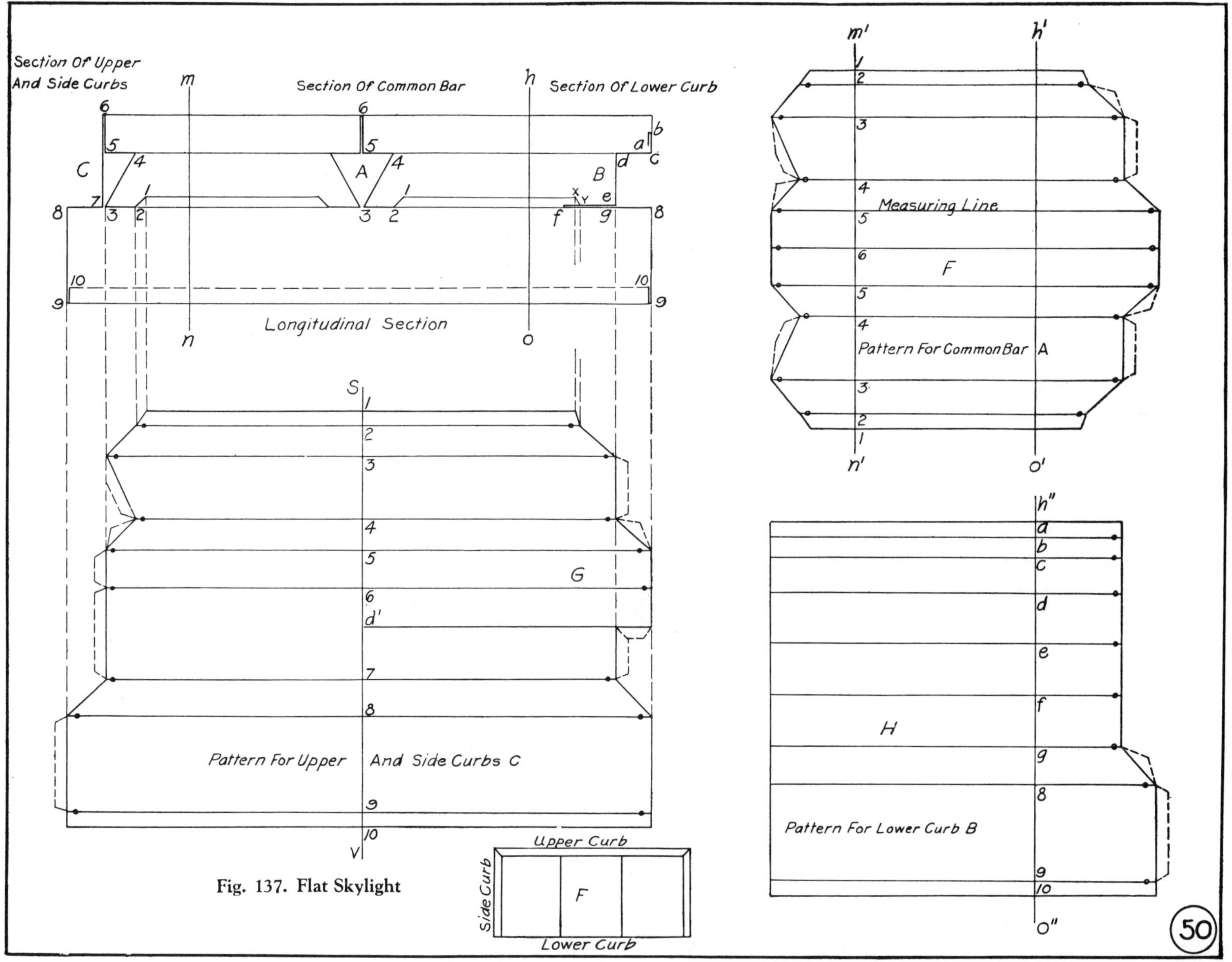

Fig. 137. Flat Skylight

pitch, it is only necessary to divide the given span or width of the skylight by the number of the pitch to get the rise or height on the center line, as shown by *g–5* in the elevation. For example, if the span of the skylight was 8 feet, a one-half pitch would rise on the center line one-half of 8, or 4 feet. Should a fourth pitch be required and the span is 8 feet, then one-fourth of 8 is 2 feet, or the rise from *g* to *5* on the center line.

Next, draw the profile of the curb *F*, placing the edge of the glass line upon the point of the triangle at *c*, and taking the dimensions from profile *F′*, make the formation of the curb, as shown. A tracing of one-half of the common bar *B′* is now placed in position upon the glass line *c–5* of the end elevation, as indicated by the figures *1, 2, 3, 4, 5* and *6* in profile *B*, which is also a section of the gable.

Take the various dimensions of the ridge bar from profile *A′* and construct the profile of this bar in its proper position, as shown at *A*. When constructing the profiles at *A* and *F*, the student must be careful to see that the rest for the glass is in the same plane as the rest for the common bar *B*.

The pattern for the gable end can be made in one piece, but it is best to make it in two pieces, as it can be cut from the metal with less waste and formed much more easily upon the brake. A one-half pattern for the gable end is shown in *G*, and is obtained as follows: First, draw the stretchout line *SV* at right angles to the glass line in the elevation, and upon this line place the stretchout of the half bar, as shown from *1* to *6* in profile *B*. Thru these points, draw the usual measuring lines, which intersect by lines drawn from similarly numbered points of intersection of the half bar with the ridge and curb, as shown. Next, from point *6′* of the pattern, draw the vertical center line *6′–x*, and upon this line locate the point *7′*, making the distance *6′–7′* equal to the distance *6–7* on the center line in the end elevation.

Now, at any convenient distance from the center lines in both pattern *G* and the end elevation, draw the vertical lines *mn*; in each case, they must be the same distance from the center line. The stretchout of the lower part of curb *F* is now placed upon the line *mn* in pattern *G* and measuring lines drawn, as shown from *7* to *12*.

Measuring in each case from the line *mn* in both elevation and pattern, take the various distances to the points on curb profile *F* and place them upon similarly numbered measuring lines in pattern *G*. Connect these points, completing the half pattern for gable.

The pattern for the miter cut on curb *F* is shown in *E*. The stretchout of profile *F* is placed upon the line *mn*, and the miter cut from *7* to *12* is obtained in precisely the same manner as the lower part of pattern *G*, the upper part of the pattern from *7* to *a* being simply a straight cut.

The pattern for the common bar is shown in *H*, and is obtained as follows: At right angles to the line *6–6″* in the end elevation, draw the line *JL*, upon which place the stretchout of the common bar, as shown by the figures *6* to *1* in profile *B*, and similar figures on the line *JL*. Thru these points, and at right angles to *JL*, draw measuring lines and intersect them by lines drawn at right angles to *6–6″* from the points of intersection on the ridge bar and curb, as shown. A line traced thru these points will be the pattern for the common bar.

Problem 102. Hipped Skylight

A hipped skylight is one that has an equal slope or pitch on all of its sides. The curb forming a continuous molding that passes horizontally around the four sides of the base is usually set on a level roof curb, for the four glazed sides provide the necessary pitch to shed the snow or rain.

Fig. 138. Double-Pitch Skylight

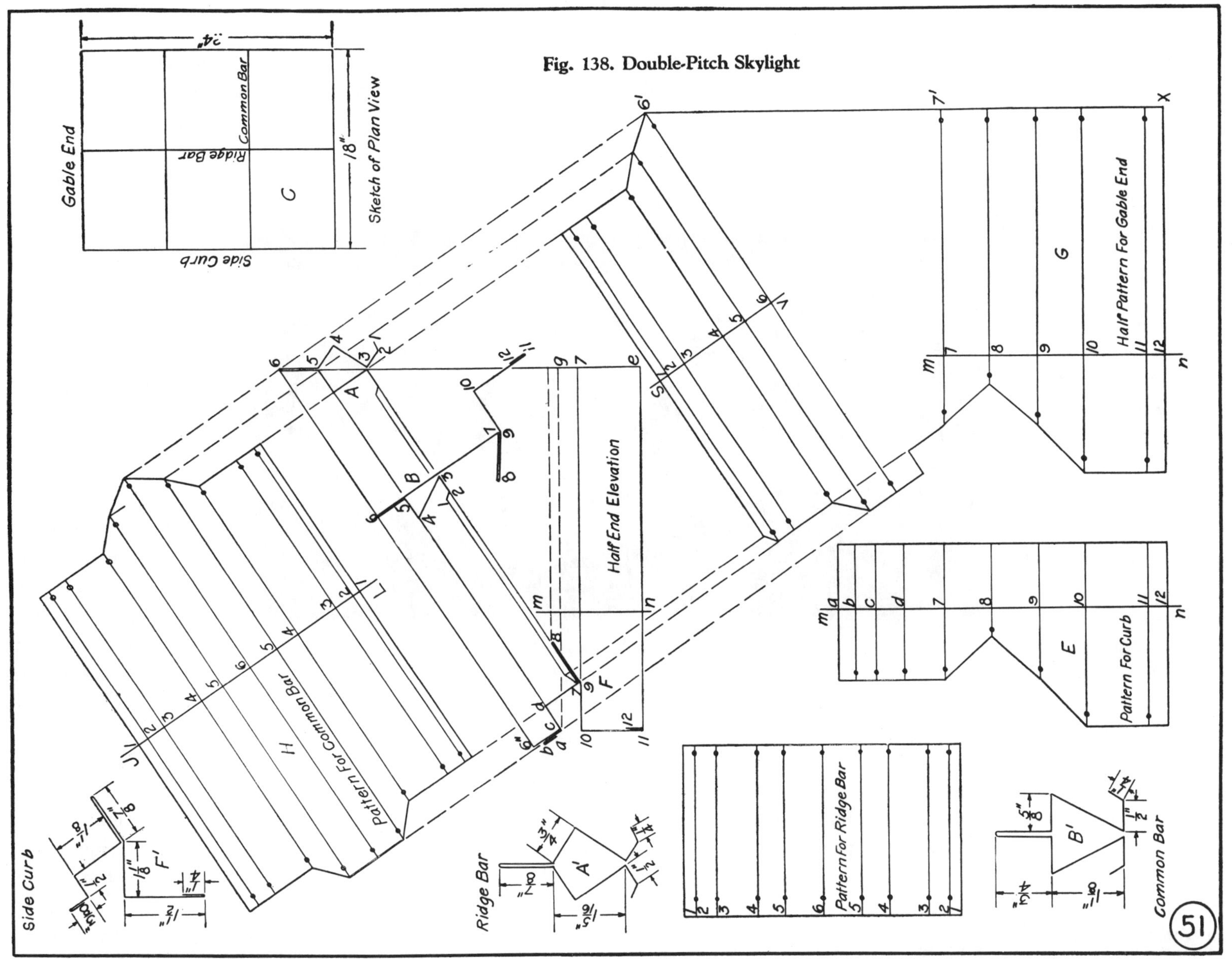

Figures 139 and 140 show the method for obtaining the patterns for a rectangular or square hipped skylight having a one-third pitch. The same principles are applicable to all hipped skylights, no matter what the pitch of the skylight may be or what angle its base may have. The drawings are one-fourth full size. When the full-size patterns are developed, they will be of such size as can be used for practical work.

In Figure 139 is shown a one-half sectional view, a quarter plan, and patterns for the curb, ridge, jack and common bars; also a diagram showing a detailed plan of intersections of the various bars used in the construction of hipped skylights.

First, draw any center line, as 9–x, at right angles to which draw the horizontal line A–2 equal to 10 inches, or one-half the width or span of the skylight. Double 10, which makes 20, and divide by 3 or the pitch required, and place this distance, which is the height or rise, upon the vertical center line from A to 2′, as shown. Now, draw the hypotenuse of the right-angle triangle 2′–2, which represents one-third pitch. At right angles to 2′–2, place a section of the common bar, as shown by A; thru this draw lines parallel to 2′–2, intersecting the curb B at the bottom and ridge bar F at the top.

Next, draw the section of curb B, placing the edge of the glass line upon the point of the triangle at 2 and the curb line gh directly below it, making the formation of the curb, as shown. A section of the ridge bar is next placed in position, as shown from 9 to 8 in F. Number the corners of the common bar A, as shown, from 1 to 6 on each side, thru which draw lines parallel to 2′–2 until they intersect the curb at the bottom and the ridge bar at the top, as shown by similar figures 1 to 6. The member 5–6, or gutter side of the bar, is cut away at any convenient distance from the end to permit of the free discharge of the water from the common bar into the gutter of the curb. This completes the one-half sectional view of the skylight, from which the pattern for the common bar can be developed as follows: At right angles to the line 2′–2, draw the line mn, upon which place the stretchout of profile A, as shown by the figures 1 to 6. Thru these points draw the usual measuring lines, and intersect them by lines drawn at right angles to 2′–2 from similarly numbered intersections on curb B and ridge bar F. Then a line traced thru the various points of intersection will be the pattern for the common bar.

The pattern for curb B is obtained by taking the stretchout of

Problem 102.   Hipped Skylight.

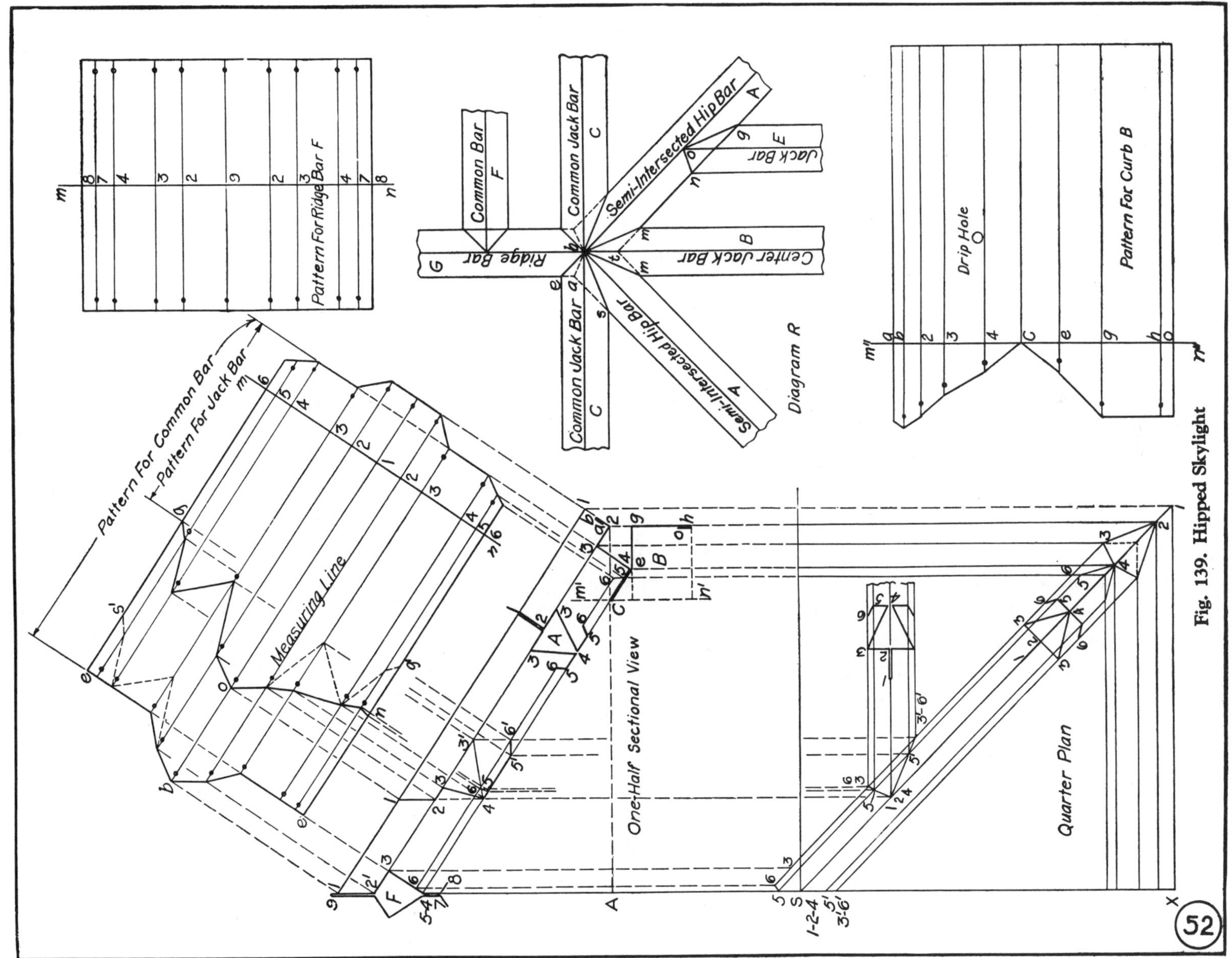

Fig. 139. Hipped Skylight

the various corners in the curb, *a, b, 2, 3, 4, C, e, g, h* and *o*, and placing them on the vertical line *m″n″* as indicated by similar letters and figures in the pattern. Thru these points draw horizontal measuring lines, as shown. Next, thru point *C* in the curb profile *B*, draw the vertical line *m′n′*. Measuring from the line *m′n′*, take the horizontal distance to the various points on curb profile *B* and place them on similarly numbered measuring lines in the pattern, measuring in each case from the stretchout line *m″n″*, as shown. A line traced thru these points will be the half pattern for the curb *A*, condensation hole to be punched into the curb between each light of glass is shown in position above line *4* in the pattern.

Jack Bar.—Before the patterns for the hip and jack bars can be developed, a quarter plan view must be drawn, which will give the points of intersection between the hip and jack bar, between the hip and ridge bar, and between the hip and curb. As the skylight forms a right angle in plan, from any point as *S* on the center line *9–x*, draw a line at an angle of 45° and intersect it by a vertical line drawn from point *1* in curb *B*, at *1* in plan. Then the diagonal line *S–1* represents the hip line in plan.

A profile of the common bar *A* is now placed on the hip line *S–1* so as to obtain the horizontal measurements, being careful to have the points *1, 2, 4* come directly on the hip line *S–1*. Thru the various points on the profile, lines are drawn parallel to the hip line *S–1*, which intersect vertical lines drawn from similarly numbered points in curb *B* and ridge bar *F* in the sectional view. A line traced thru these points, as shown from *1* to *6* in plan, will represent the intersections between the hip bar and curb and ridge bar.

Before the pattern for the jack bar can be developed, the miter line between the jack bar and hip bar must be obtained, both in the elevation and plan, in the following manner: Take a tracing of the common bar, profile *A*, and place it in a horizontal position in plan, as shown. Thru the various points in the profile draw horizontal lines, intersecting similarly numbered lines on one side of the hip bar in plan, thus forming the miter line of the short cut from *1* to *6*; also the miter line of the long cut from *1* to *6′*. From these intersections at right angles to the lines of the jack bar, draw vertical lines intersecting similarly numbered lines in the half sectional view, obtaining the points of intersection and miter line of the short cut, shown by *1, 2, 3, 4, 5* and *6*, and the long cut, shown by *1, 2, 3′, 4, 5′* and *6′*.

For the pattern for the upper cut of the jack bar, the same stretchout can be used as that used for the common bar. Therefore, from the various intersections on the miter line in the sectional view and at right angles to the glass line *2′–2*, draw lines intersecting similarly numbered lines in the pattern for the common bar. A line traced thru these points of intersection will be the pattern for the upper cut of the jack bar, the lower cut being the same as that shown on the pattern for the common bar.

Ridge Bar.—The pattern for the ridge bar is simply a piece of metal, rectangular in form, its width being equal to the stretchout of profile *F*, as shown from *8* to *8* on the line *mn* in the pattern. The length of the ridge bar is equal to the difference between the length and width of the curb of the skylight.

Hip Bar.—Before the profile and pattern for the hip bar can be developed, a true elevation of the hip bar must be constructed, as shown in Figure 140, the one-half sectional view and quarter plan being a duplicate of Figure 139. The true elevation of the hip bar could be drawn in Figure 139, but in this case, for want of space and to save overlapping of lines, it has been transferred to Figure 140, as shown.

The true intersections between the hip bar and curb and ridge bar being shown in their proper position in the quarter plan, the

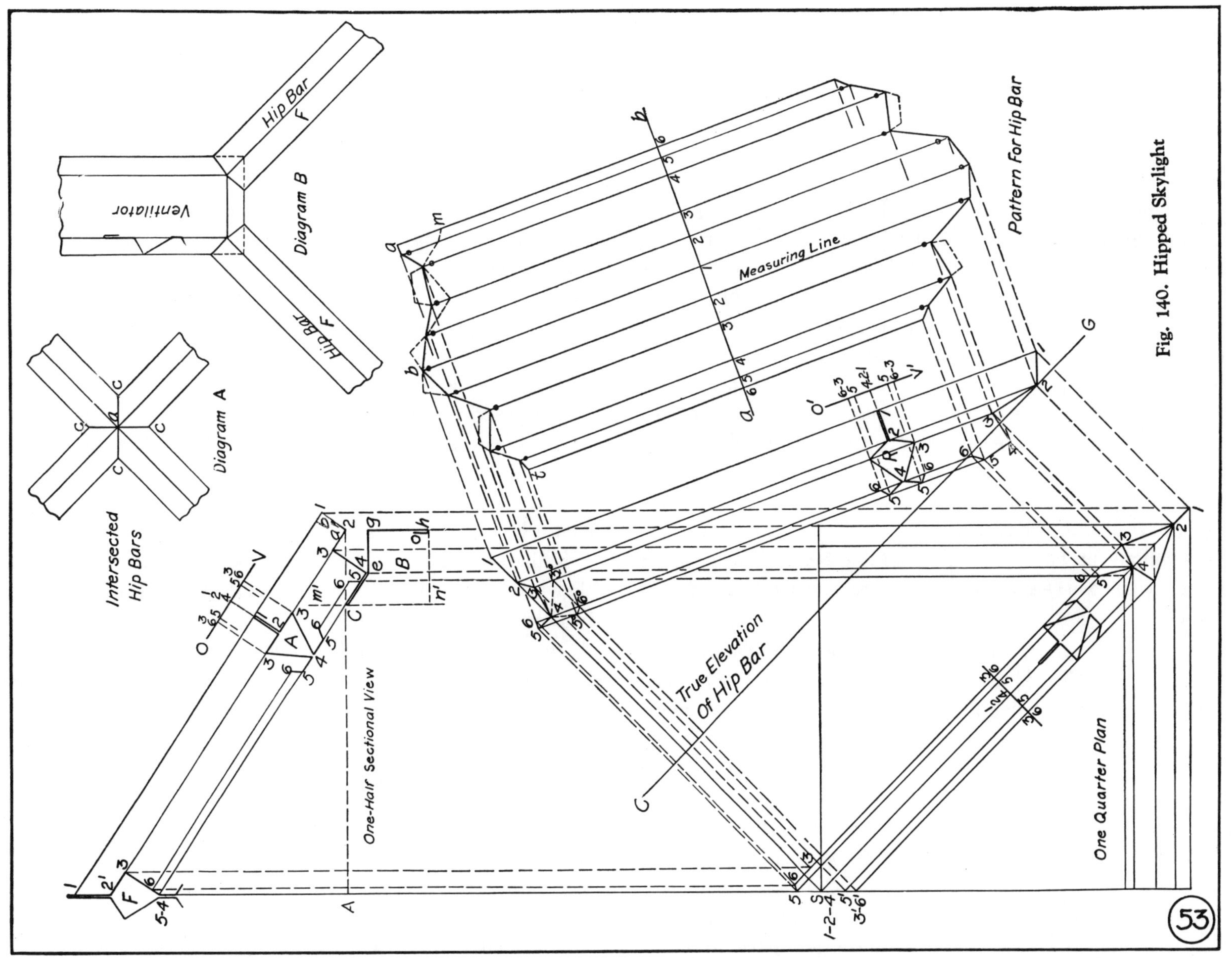

Fig. 140. Hipped Skylight

next step is to draw a true elevation of the hip bar, from which a true profile of the hip bar and pattern are obtained as follows: Parallel to the hip line *S–1* in plan, draw any line, as *CG*. From the various intersections *1* to *6* in both curb and ridge in plan, draw lines at right angles to *S–1*, crossing the line *CG* indefinitely. Since the several points in the upper miter of the hip bar must appear in the true elevation of the bar at the same heights as corresponding points of the common bar in the one-half sectional view, these distances may be transferred from the sectional view to the true elevation of the hip bar by means of the dividers. Now, measuring in each and every instance from the line *A–2* in the one-half sectional view, take the vertical heights to points *1* to *6* in the ridge and *1* to *6* in the curb, and place them in the elevation of the hip bar on similarly numbered lines drawn from the plan, measuring in each instance from the line *CG*. Thru these points, draw the miter lines and connect similarly numbered points at the top and bottom by lines, as shown. If the miter lines at the curb and ridge in the elevation are true, these lines will all run parallel to the glass line *2–2*, as shown, which completes the true elevation of the hip bar, intersecting the ridge and curb.

It is next necessary to determine the profile of the hip bar, shown at *R*. This is accomplished by drawing the line *OV* parallel to the glass line *2'–2* in the one-half sectional view. From the various points on profile *A* and at right angles to *OV*, draw lines to intersect the line *OV*, which will show the widths of the various parts of the bar, as shown by the figures *1* to *6* on the line *OV*. Now, take the various divisions on this line and place them upon the line *O'V'*, which is drawn parallel to the lines of the hip bar in the true elevation. From the various points on this line at right angles to the line *2–2*, draw lines intersecting similarly numbered lines in the elevation of the hip bar, as shown

from *1* to *6*. Connect these points on either side, as shown; then *R* will be the true profile of the hip bar.

Pattern for Hip Bar.—The pattern for the hip bar is developed by taking the stretchout of profile *R* and placing it on the line *ab* drawn at right angles to the lines of the hip bar. Thru the various divisions on the stretchout line *ab*, draw the usual measuring lines, which intersect by lines drawn at right angles to the line *2–2* in the elevation from similarly numbered points in the miter lines of the ridge at the top and the curb at the bottom. A line traced thru these points will be the pattern for the hip bar.

Bar Intersections.—The intersections of the various bars with the ridge as they would appear in the plan view of a hipped skylight, are shown in diagram *R*, Figure 139. The ridge bar *G* is intersected by the common bar *F*, the miter cut being shown by *ebe* in the pattern for the common bar. The jack bar *E* intersects one side of hip bar *A*, shown by the short miter line *no* and the long miter *og*. The long and short miter cuts are shown by similar letters in the pattern for the jack bar.

If the centers of the common jack bars *C* are mitered, as shown at *b*, one side of the bar will intersect the ridge, as shown by the miter line *be*, while the other side will intersect the hip bar *A*, shown by the miter line *bs*; then the miter cut *eb* would be the same as the cut on one-half of the common bar, and the miter cut *bs* the same as the cut on the long side of the jack bar. Then, the pattern for common jack bar *C* is simply a common bar pattern having the cut for the long side of a jack bar placed upon one-half of the pattern, as shown by the dotted lines *bs'* on the upper end of pattern for the common bar.

The two hip bars *A* and *A* in diagram *R* intersect the ridge bar, as shown by the dotted miter line *ab*, the inner half of the bars intersecting on the line *bt*. The miter cuts *tb* and *ba* are shown by similar letters on the upper end of the pattern for the hip bar. The center jack bar B intersects the two hip bars A and A on the miter lines bm, and the pattern is simply a jack bar pattern having a long cut on each side of the center line, as shown by og and Og' in the pattern for the jack bar.

Square hipped skylights are often required in practical work, and the intersections of the four hip bars are shown by the miter lines ac in diagram A, Figure 140. The miter cuts ac on each side of the intersected hip bars are shown by tbm on the upper end of the pattern for the hip bar.

Diagram B shows the intersection of hip bars F and F with a ridge ventilator, the profile of the lower part of the ventilator being the same as one-half of ridge bar F. The hip bar pattern with the miter cut ba placed upon each side of the center line, will be the pattern for the hip bar F in diagram B.

Measurements.—Having now developed the patterns for the various parts of the hipped skylight, it is usual in actual shop

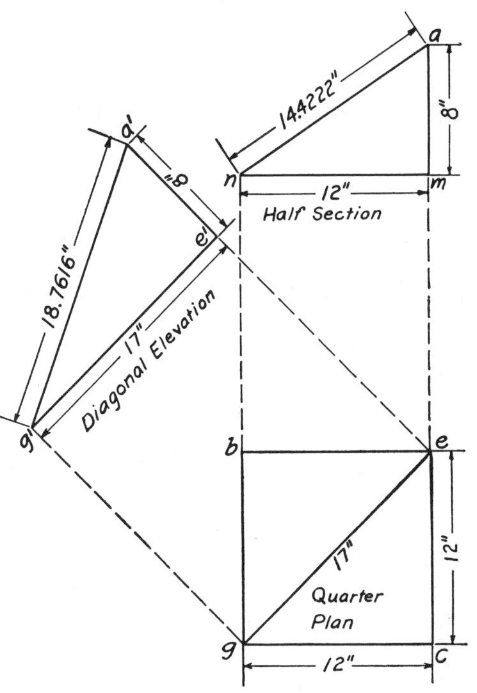

Fig. 141. Rule for Finding Factors

practice to transfer the various patterns from the drawing to sheet metal by means of a prick punch, marking the measuring points and the name of the pattern upon each one. The metal patterns are kept for future use and can be used when constructing any style hipped skylight having a one-third pitch, no matter what the size of the curb may be.

The size of the curb, also the widths of the ventilators, when used, forms the basis for obtaining the lengths of the various parts of a hipped skylight. There are several methods used for finding the true lengths of the ventilator, common, jack, and hip bars; but the best results are obtained by mensuration, in which a factor or multiplier is used to find the length of the bars without using any drawings, diagrams, scales, triangles, etc., when the size of the curb and width of the ventilator are known.

The following rules should be observed for finding the true lengths of the various parts of a hipped skylight by mensuration:

1) To find the true length of the ridge bar, deduct the shortest side of the curb from the longest side.

2) To find the true length of the ventilator, deduct the shortest side of the curb from the longest side and add the width of the ventilator.

3) To find the length of the common bars for a one-third pitch skylight, divide the short side of the curb by 2 and multiply by the factor 1.2.

4) The distance on the curb between the hip and jack bars, multiplied by the factor 1.2, will give the true length of jack bars for a one-third pitch skylight.

5) To find the true length of the hip bars for a skylight having a one-third pitch, divide the short side of the curb by 2 and multiply by factor 1.56.

6) To find the true lengths of the common, hip and jack bars in a hipped ventilating ridge skylight, deduct the width of the

ventilator from the short side of the curb, then divide by 2 and multiply by the same factors used for one-third pitch skylights.

The factors here given can be used only for skylights having one-third pitch; also the curb line must always run in line with the edge of the glass line, as shown by *2*, *g*, *h*, in Figure 140.

The rule for finding the factors used for one-third pitch is shown in Figure 141. If the rise of the common bar, shown by *am*, is 8 inches, and the base *mn* is 12 inches, the length of the hypotenuse *an* will equal the square root of the sum of the squares of the rise and base, or 14.4222 inches. Now, divide this length 14.4222 by 12, the length of the base, which gives 1.2018. This number and decimal will be the factor to use for finding the length of the common and jack bars. The factor for the hip bar is found in a similar manner. As the length of the hip bar, *eg* in plan, equals the square root of the sum of the squares of the two sides *ec* and *cg*, the true length of the hip bar shown by *a'g'* in the diagonal elevation, will equal the square root of the sum of the squares of the two 12-inch sides and the square of the 8-inch rise, shown by *e'a'*, thus: $\sqrt{12^2 + 12^2} = \sqrt{288 + 64} = \sqrt{352} = 18.7616$ inches.

Now, divide 18.7616 by 12 and the quotient 1.5634 will be the factor to use in finding the length of the hip bars. In practice, 1.56 is used. Should the skylight have a different pitch or rise from 8 inches to 1 foot of base, the factors can be found in a similar manner.

Problem 103. Rectangular Flaring Pan with Folded Corners

The sheet-metal worker is often required to construct square and rectangular flaring pans, the corners of which are to be made water-tight by folding together and turning them to the sides or ends of the pan, having the upper edges of the folded parts finish neatly under the wire which is enclosed along the top.

A pan constructed in this manner is water-tight without soldering, and, since this style of pan is often subjected to a temperature higher than the melting point of solder, the construction shown in Figure 142 is important. In actual shop practice, the pattern is laid out with a steel square and the dividers directly upon the sheet of metal.

In Figure 142 is shown the full pattern and side elevation of a rectangular flaring pan having an equal flare on all sides. Draw the side elevation, shown by $ACBG$, the lines AB and CG show the slant height of the pan. Next, draw the plan of bottom $EHDF$. From each corner of the plan extend the lines, as Hn, En, etc., in each case making the distance from the corner to n equal to the slant height of the sides, shown by GC in the side elevation. Now, thru these

Problem 103. Rectangular Flaring Pan with Folded Corners.

points parallel to EH and EF, draw lines as am, which intersect by vertical lines drawn from A and C in the side elevation. Connect the corners E to a and H to m. Since the flare of all sides of the pan are the same, as shown by an and nm, place these distances upon the line nm at each of the four corners and draw the miter lines, as shown. Then will $amstxdfb$ represent the pattern for the pan if the corners were butted together. The allowance for wire is now added, as shown by the dotted lines.

The next step is to provide for the extra material required for a folded corner, and the manner in which this result is accomplished is shown in the upper left-hand corner of the pattern. Bisect the angle Eab, obtaining the line of bisection RE. Set the compasses to a radius slightly less than the distance from the point a to the line RE, and from a as center, describe the short arc oc, intersecting the flare line aE at g. With g as center and gc as radius, intersect the arc oc at e. Draw a line from a thru e, producing the line until it meets RE in the point h. From h draw a line to b. Then $Rbha$ is the amount to be notched from the corners, so that when folded, they will finish neatly under the wired edge of the pan.

Problem 104. Round Ventilator with Molded Base

This problem is presented as an example of the practical application of the three methods of development employed in developing patterns for previous problems in this course. Since the drawings represented are those commonly used in a sheet-metal shop, the student is expected to construct the various patterns, aided only by the instructions he has already received.

Figure 143 shows the elevation of a cupola and ventilator with a square molded base that fits over a double-pitched roof. The drawing is shown one-fourth full size, and the different parts of the article are indicated by letters, as shown. The lower portion of the ventilator, shown in F, is square in form and fits over a roof having a one-third pitch. A tapering transition piece from square to round, shown in A, is required between the square base and the round pipe, shown in section G. Section H is the frustum of a cone whose sides flare at an angle of $45°$ with the horizontal, and which is placed over the upper edge of the round pipe, as shown. The angle of inclination of upper section C is $47\frac{1}{2}°$ to the horizontal, and is joined at the top by the round tapering piece B and the ball E. A circular band or windshield is shown by $acbe$ in the elevation. The band is strengthened by enclosing a wire

Problem 104. Round Ventilator with Molded Base.

in its upper and lower edges, and is fastened to the upper flange C and lower flange H by means of a band iron brace, as shown in the drawing. Draw the full size elevation, and, after the patterns have been developed for the various sections, the customary allowances for laps or riveting edges should be added to the drawing.

Problem 105. Ball, Development by Zones

This problem is illustrated by Figure 144. The sphere is the most prominent example of the many forms whose surfaces admit of approximate development. Since solids of this kind may be resolved into parts resembling those capable of accurate development, the same general methods are applicable. Thus, in the case of the ball, shown in Figure 144, this solid has been resolved into a number of frustums of cones, and patterns developed in the regular way. This method is called development by zones or horizontal sections, and may be used for obtaining patterns or blanks for any curved work, whether hammered by hand or machine.

The ball can also be made from gores or vertical sections, but many sheet-metal workers prefer to make them from zones, as the pieces are not so apt to warp, and the labor is less in raising the zones than in raising the gores.

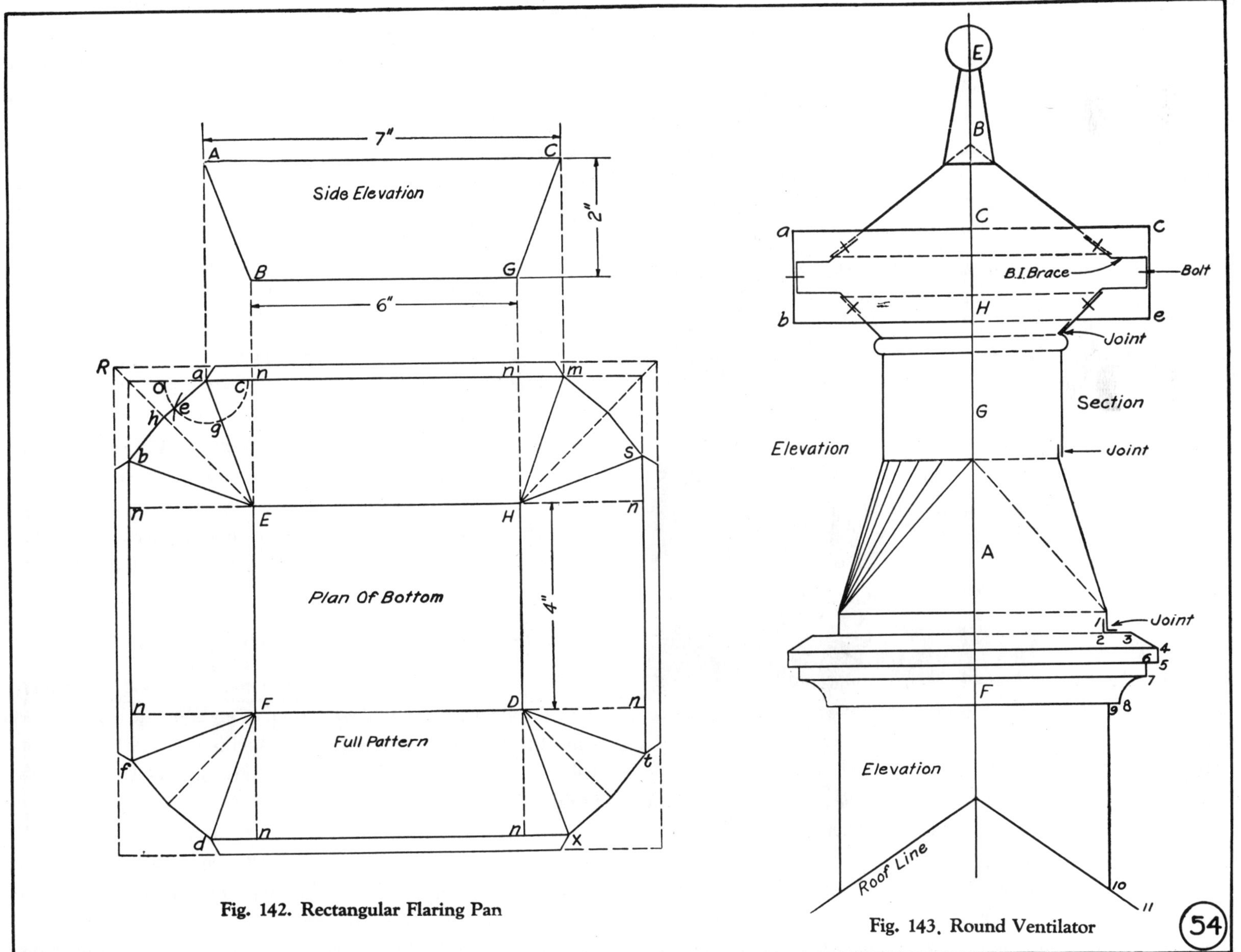

Fig. 142. Rectangular Flaring Pan

Fig. 143. Round Ventilator

In Figure 144 is shown the elevation and patterns for a ball constructed by means of zones. First, draw the elevation of the ball the required size and divide into as many zones as desired— in this case, six. When the ball is large, more zones must be used so as to require the least amount of labor in raising the blanks to form, for it takes less time to cut an extra blank and raise the ball to a truer circle than to use less zones and raise them to a greater depth.

Having divided the quarter circle into three equal parts, shown by *abeg*, the radii with which to develop the patterns for zones *A*, *B* and *C* are obtained as follows: Thru the center of the ball,

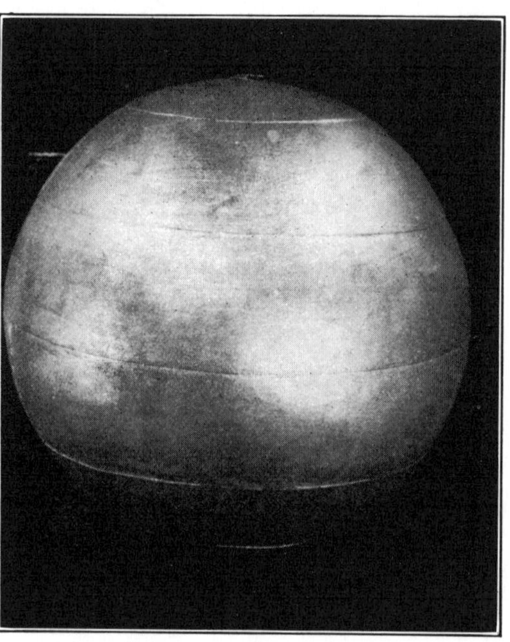

Problem 105. Ball, Development by Zones.

draw the vertical line *F H*; then, thru the extreme points of the zones *A* and *B*, draw the lines *ge* and *eb*, which extend until they meet the center line at *F* and *G*, respectively. Where the two zones *A* and *B* joint together on the line *e1*, use *m* as a center and describe the quarter circle *1–5–m*, which represents the half section thru *1–m*. Divide the half section into a number of equal spaces, as shown from *1* to *5*. With *F* as center and radii equal

to *Fe* and *Fg*, describe the arcs *e'f* and *g'h*. Now, draw any radial line, as *g'f*, and, starting from *e'* on the upper arc, step off twice the number of spaces contained in the section *1–m*, as shown from *1* to *5* to *1* in the pattern. From *F* thru *f*, draw a line intersecting the arc *g'h* at *h*, completing the half pattern for zone *A*, to which laps are allowed, as shown. To obtain the pattern for zone *B*, use *n* as center, and with radii equal to *Gb* and *Ge* in the elevation, describe the arcs *b'* and *cc'*. Draw the radial line *cn*, and, starting from *c*, step off on the center arc four times the number of spaces contained in the section *1–m*, and draw a line from *c'* to *n*, intersecting the inner arc at *b'*. Then *cb''c'b'* is the full pattern for zone *B*. The upper zone *C* is formed of a circular disc of metal; the pattern is described with a radius equal to *ab* in the elevation. When constructing the drawings, no attention is paid to the lower half of the ball, since the patterns for the upper half will serve, also, for the lower half. After the patterns have been formed to a true circle and soldered together on the inside, they are subjected to the raising process, and the desired curve is made with the raising hammer upon a wood or lead block.

Problem 106. Ball. Development by Gores

In Figure 145 is shown the approximate method for developing the patterns for a ball composed of eight vertical sections, as shown in the plan view. This method is referred to as the development by gores. The patterns are obtained by means of parallel lines as follows:

Let *B* represent the plan view of the ball, which is divided into as many parts as gore pieces required. In this case, the ball is to have eight gore pieces, altho any number of pieces can be used, and the principles will apply. With *1'–m'* as radius and *m* as center, describe the quarter circle *m–1–7*, which represents

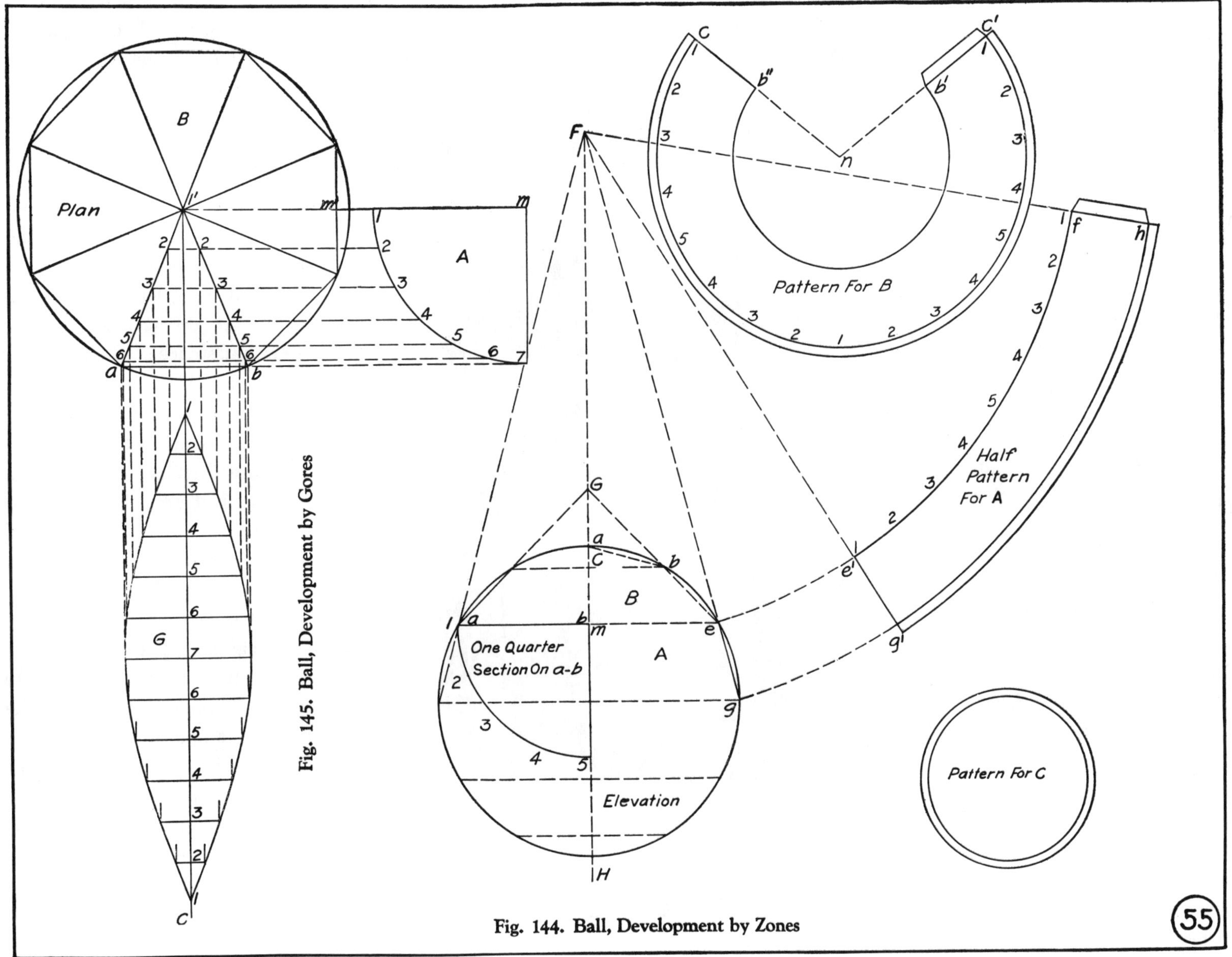

Plan
B
A
Pattern For B
Half Pattern For A
Fig. 145. Ball, Development by Gores
G
One Quarter Section On a-b
B
A
Elevation
Pattern For C
Fig. 144. Ball, Development by Zones

the half section thru *1'–m'*, shown in *A*. Divide the half section into a number of equal spaces, as shown from *1* to *7*, and from these points draw horizontal lines, intersecting the miter lines *1'–a* and *1'–b* in plan, as shown. At right angles to *ab* draw the center line *1'–c*, upon which place twice the number of spaces contained in half section *A*, as shown from *1* to *7* to *1* in pattern *G*. Thru these points draw lines at right angles to *1'–c*, which intersect by vertical lines drawn from similarly numbered points on the miter lines *1'–a* and *1'–b*. A line traced thru these intersections, as shown in *G*, will be the desired pattern. A small edge should be added to one side of the pattern, so that the gores can be joined together by means of a lapped seam.

Problem 107. Base for Chimney Top, Rectangular to Round

In Figure 146 is shown a short method for laying out the pattern for the base of a chimney top in the form of a transition piece from rectangular to round, which is to be constructed in two pieces. In actual shop practice, the pattern is laid out with the steel square directly upon the sheet of metal. Let *ABCD* represent the portion of a sheet of iron, which can be of any desired width. The rectangular base in this case is assumed to be 13 inches by 17 inches, and the diameter of the round top 7 inches. About 5

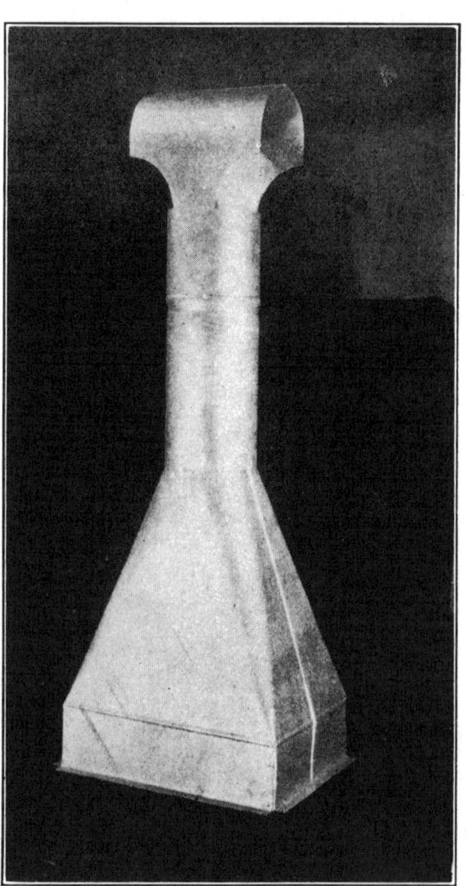

Problems 107 and 108. Base for Chimney Top and Hood.

inches from the bottom of the sheet, draw the line *mn* parallel to *BD*, and upon this line locate point *a*, making the distance *ma* a little more than one-half of the width of the short side of the base. Next, measuring from *a*, make *ab* equal to the length of the long side of the base; in this case, 17 inches. Bisect *ab* and erect the perpendicular, as shown by the center line *ge*. Now, measuring from point *e* on the upper edge of the sheet of iron, locate the points *o* and *h*, making the distance *oh* equal to one-half the circumference of the top, after which the outer edge of the steel square is placed upon the points *o* and *a*, being careful to have the 6-1/2-inch mark on the short arm of the square directly over point *a*. Lines are then scribed upon the metal, as shown by *ot* and *ta*. Now, move the steel square along the line *ot* to *x*, making the distance *tx* equal to *m B*, which is the width of the vertical flange that extends down the side of the chimney. Then draw the lines *txv*. After this, reverse the square to the position, shown in *S*, to obtain the cut for the corners *b* and *a*. The opposite side of the pattern is treated in the same manner, and laps are added, as shown by the dotted lines. The circular cut on the upper edge of the pattern is obtained by extending the line *xo* and *x'h* until they intersect at *F*. With *F* as center, and *Fo* as radius, describe an arc, which will be the upper edge of the pattern. This completes the half pattern for a

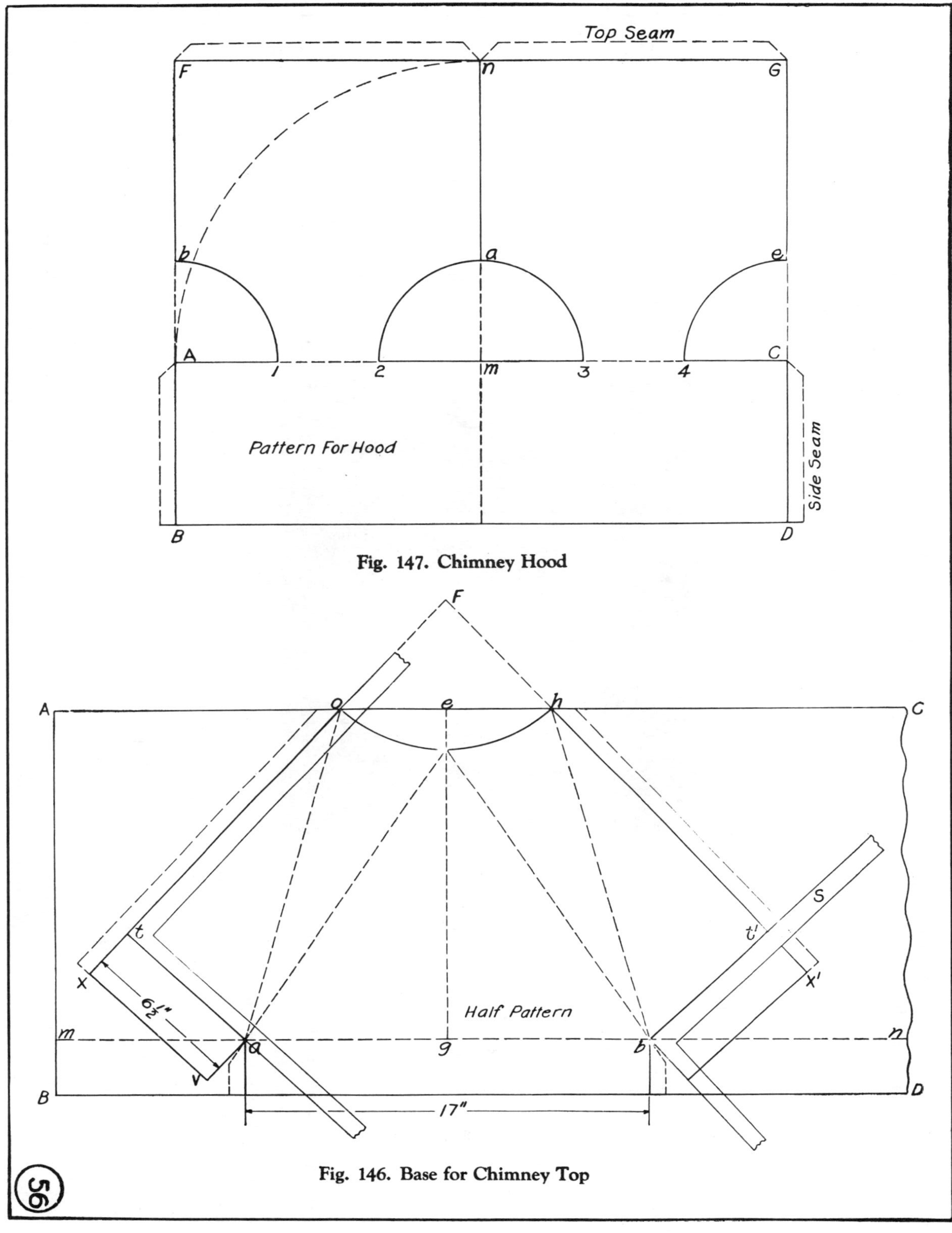

Fig. 147. Chimney Hood

Fig. 146. Base for Chimney Top

chimney base, rectangular to round, made in two pieces; the other piece can be cut from the same sheet of metal without any waste by simply reversing the position of this pattern when placing it upon the sheet.

Problem 108.   Chimney Hood

Figure 147 shows the pattern for a one-piece chimney cap which can be laid out directly upon the metal. First, lay out the pattern for the round pipe of the width and circumference required, as shown by *ABCD*. Bisect the line *AC* and erect the perpendicular *mn*. With *m* as center and *mA* as radius, describe the arc *An*. Thru point *n* draw a horizontal line parallel to *AB*, which intersect by the vertical lines *BA* and *DC* extended to *F* and *G*. Next, divide the two spaces *Am* and *Cm* into three equal parts, as shown. With the dividers set equal to one of these spaces, as *m–2*, and using points *CA* and *m* as centers, describe the half circle *2–a–3* from *m*, and the quarter circles *b–1* and *4–e* from *A* and *C*, as shown. Cut down the center line from *n* to *m*: then cut out the half circle *2–a–3* and the quarter circles *A–b–1* and *C–4–e*. Laps are allowed for seaming on the top and sides, as shown. Form up the pipe in the forming rolls, and groove the edges *AB* and *CD*. Then form the top and make a grooved or riveted seam on the edges *Gn* and *nF*. This cap can be quickly made without waste of material, and is frequently used as a cure for

chimneys that are troubled from the rebound of the wind against a higher building. When used for this purpose, the side of the cap is set parallel with the side of the building.

Problem 109.   Florist's Watering Pot

In Figure 148 is shown the elevation and patterns for the various parts of a florist's watering pot. The body of the sprinkler, shown at *A*, has a wire enclosed in the top at *mn*, the lower edge being connected to the bottom *fv* by means of a double seam. At *F* is shown the tapering spout which joins the body at *s* and *t*. The sprinkler rose *R* is soldered to the tapering tube or socket *H*, which slips down over the end of the spout *F* to hold the sprinkler rose in position, and also allows it to be removed for cleaning and other purposes. The circular handle *B* with its hand grasp or boss on the inner side, shown at *h*, is placed over the center of the top; the ends being attached to the body in the front and rear, as shown at *m* and *g*.

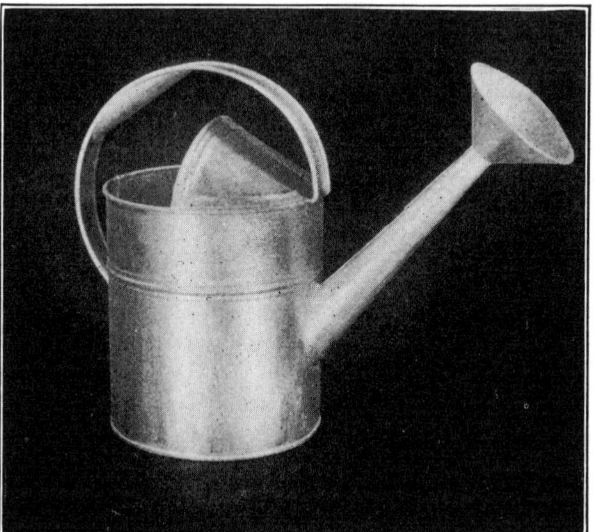

Problem 109.   Florist's Watering Pot.

The helmet lip or breast *C* has a wired edge, shown by *db*, the lower edge *mb* is soldered to the wired edge of the body on the line *mn*, as shown. This is the most common form of breast which is added to certain vessels for liquids, particularly such as are used for pouring.

The pattern for the breast is shown in *C′* and is developed by the parallel-line method as follows: With *b* as center and *bm* as radius, describe the quarter circle *me*, which will represent a one-

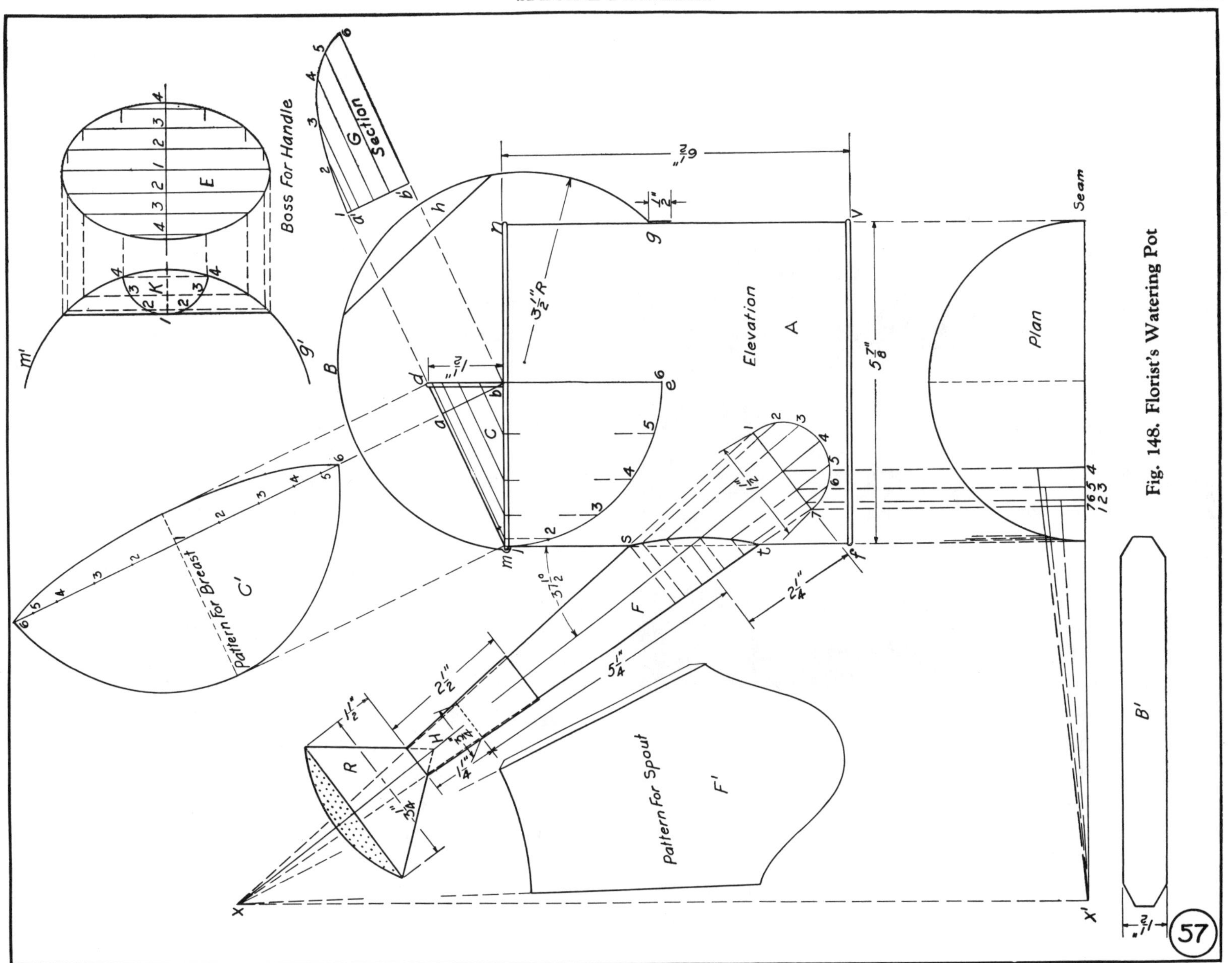

Fig. 148. Florist's Watering Pot

half plan of the breast. Divide the quarter circle *me* into a convenient number of equal spaces, as at *1, 2, 3, 4, 5* and *6*. From these points erect vertical lines, which intersect the base line *bm*, as shown. Now, from the points on the base line *mb*, parallel to the line *md*, draw lines indefinitely intersecting the vertical line *db*, as shown. Draw the perpendicular line *a′b′*; then, measuring from the line *mb*, take the various distances to the points *1, 2, 3*, etc., on the quarter circle *me*, and place them on similarly numbered parallel lines extended in section *G*, as shown. A line traced thru these points will complete the full view of a cross section of the breast, as it would appear if taken on the line *ab* in the elevation. The pattern is now ready for development by the parallel method.

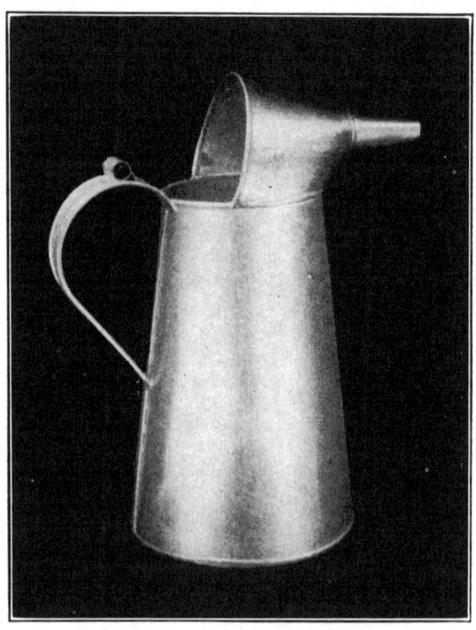

Problem 110.   Automobile Measure.

From point *b* at right angles to *md*, draw the stretchout line *b–6*, upon which place twice the number of spaces contained in section *G*, as shown from *6* to *1* to *6*. Thru these points, draw the usual measuring lines, which are intersected by like numbered lines drawn at a right angle to the line *md* from similarly numbered points on the lines *mb* and *bd*. A line traced thru these intersections will be the pattern for the breast, as shown in *C*.

The outline of the tapering spout is shown in *F*. Complete the full cone in the elevation, as shown by the dotted lines, and produce them to the apex at *x*. Next, draw the plan and develop the pattern for the spout, shown at *F*, by the method described in Problem 63.

The sprinkler rose *R* and the tapering socket *H* are the frustums of two cones, and the patterns are developed by the radial method described in Problem 43. The pattern for the circular handle *B* is shown at *B′*, and is a strip of metal equal in length to *mBg* in the elevation.

At *E* is shown the method of laying out the pattern for the boss on the inner side of the handle, shown at *h* in the elevation. This piece is added so that the handle may more readily be grasped and held firmly in the hand. Take a tracing of the boss and place it in position on the arc *m″g*, as shown. The curved outline, shown in *K*, is a section thru the central portion of the handle. Divide the section into equal spaces, as shown from *1* to *4*. Thru these points, draw vertical lines, intersecting the outline of the handle on the arc *m′g′*, as shown. The girth of section *K* is then placed upon the stretchout line, and the pattern is obtained by the method of parallel developments, shown at *E*.

Problem 110.   Automobile Measure

Figure 149 shows the method of developing the patterns for a liquid measure or pouring-can having an offset funnel attached to the top. First, draw the elevation of the flaring measure with the funnel attached, in accordance with the dimensions given in the drawing. The patterns for the body of the measure and the spout for the funnel, shown by *a–b–1–7*, are developed by the radial method and no further explanation is necessary. The pattern for the spout is shown at *R*.

The pattern for the funnel *A*, shown by *1, 7, 8, 12′, 16′*, is de-

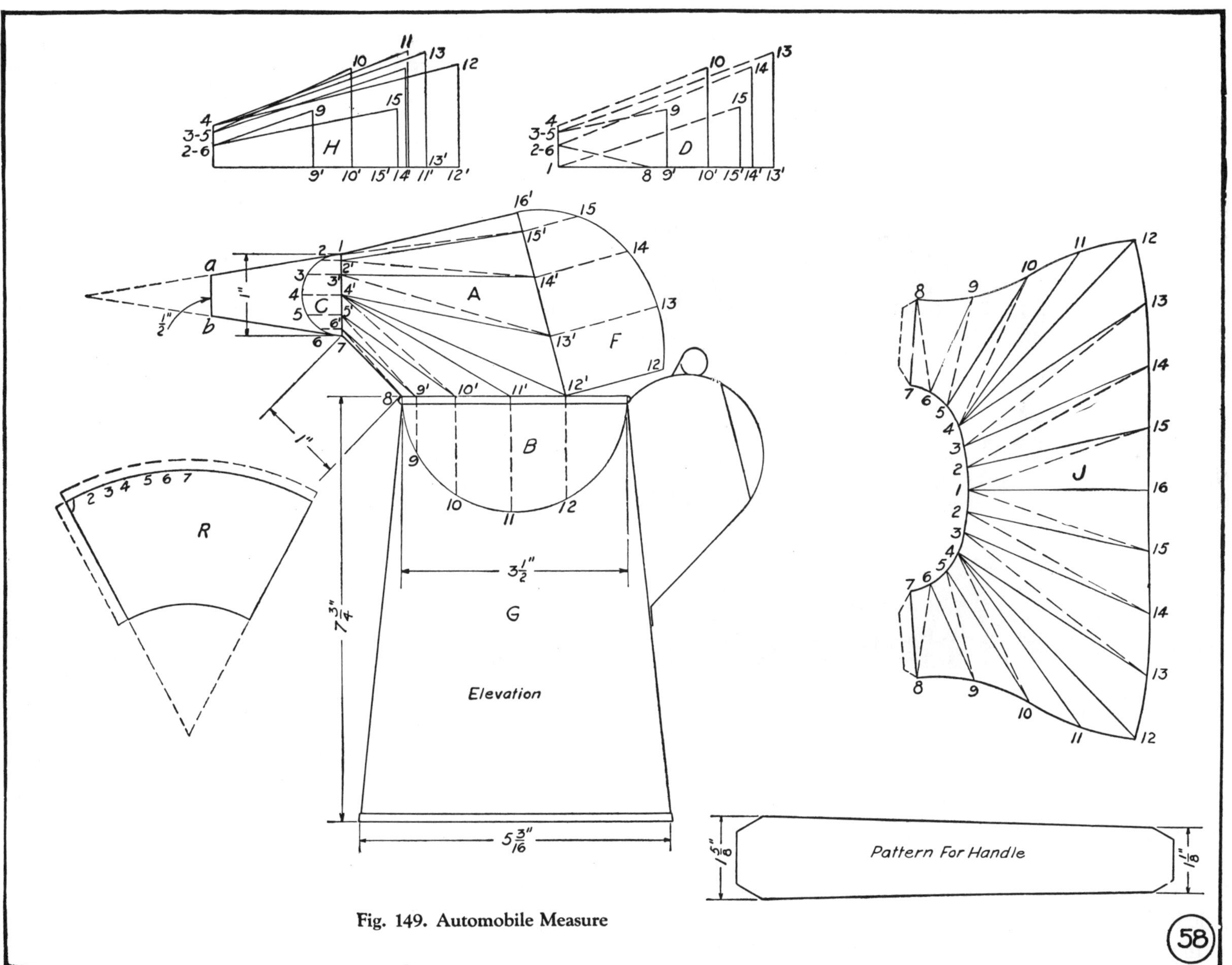

Fig. 149. Automobile Measure

58

veloped by the short method of triangulation described in Chapter **VIII**. Construct half sections on the funnel, as shown by *CB* and *F*. These half sections represent the shape of the funnel on their respective lines. Divide each section into equal spaces, and from these points draw perpendicular lines, which intersect the base lines *1–7*, *8–12'*, and *16'–12'*, as shown. Then connect the various points on the base lines with the usual solid and dotted lines, shown in *A*. The true lengths of the solid and dotted lines are shown in diagrams *H* and *D*, and are found in the customary manner, as described in Problem 80.

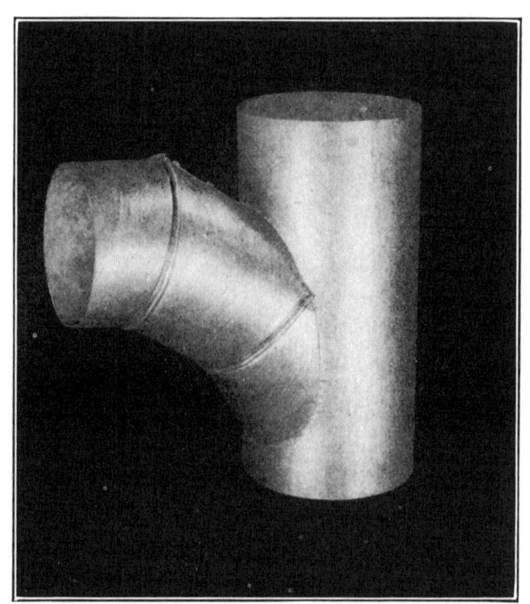
Problem 111.   A Pieced Elbow Intersecting
a Round Pipe.

The full pattern for the funnel is shown in *J*, and is developed in precisely the same manner as described in connection with Figure 124. The seam line *7–8* and the center line *1–16* in the pattern are equal to *7–8* and *1–16'* in the elevation. The divisions from *1* to *7* in the pattern are equal to the divisions from *1* to *7* on half-section *C*. The divisions *8* to *12* on the lower edge of the pattern are equal to the divisions *8* to *12* in half-section *B*. The

divisions from *12* to *16* on the upper edge are equal to the spaces from *12* to *16'* on the outline of half-section *F*. After the pattern is constructed, edges are added for seaming and wiring.

Problem 111.   A Pieced Elbow Intersecting a Round Pipe

In the sheet-metal worker's experience, it is often necessary to connect two pipes of different diameters whose axes lie at right angles to each other. When making a connection of this kind in blow-pipe work, it is important that the fitting be so constructed as to secure an easy flow of air thru the pipes. A fitting that will accomplish this result is shown in Figure 150.

The vertical pipe in this case is assumed to be 6 inches in diameter. It is connected to a horizontal pipe 4-3/4 inches in diameter by means of a four-piece elbow, shown by *K HFB*. First, draw the angle *agb* and make it equal to that of the required fitting; that is, 90°. Next, on the line *gb* lay off a distance of 5 inches from *g* to *e*, which is the radius desired for the inner curve of the elbow. Make *eb* equal 4-3/4 inches, the diameter of the elbow, and, with *g* as center and *ge* as radius, describe an arc and construct the full elevation of the elbow, as shown. Since this process has been fully described in Problem 16, the instructions need not be repeated here. After the elevation of the elbow has been drawn, an elevation of the vertical pipe is placed in position, as shown by *ABCD*. Draw the profile of the elbow, as shown at *G*, dividing it into equal parts, as shown from *1* to *7*. Draw a plan view of the vertical pipe, shown at *R*, and place duplicate of profile *G* in elevation in its proper position, as shown by *G'* in plan. From the points *1* to *7* in profile *G'*, draw horizontal lines to the right until they intersect the outline of the larger pipe, from which intersections erect vertical lines indefinitely. From the various points in profile *G* in elevation, draw

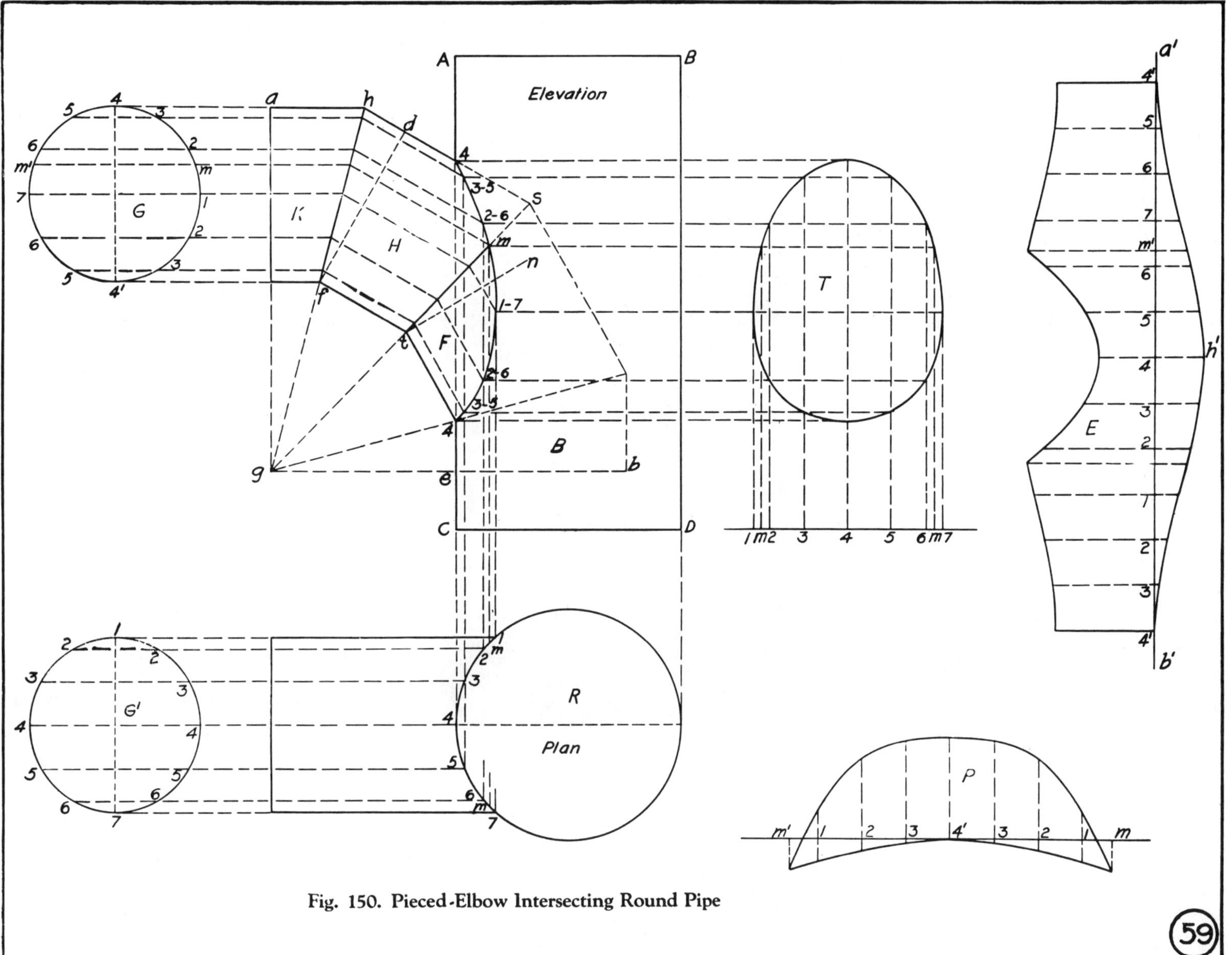

Fig. 150. Pieced-Elbow Intersecting Round Pipe

horizontal lines until they intersect the miter line *fh*, afterwards producing them in the manner shown in *H* and *F* until they intersect corresponding vertical lines previously drawn from the plan. A line traced thru these intersections will represent the miter line between the vertical pipe and the sections *H* and *F* of the elbow. This irregular curve crosses the miter line *st* at *m*, and this point is projected to profile *G* in the elevation and profile *R* in the plan, as shown at *m* and *m'*. These points will be used in developing the patterns. It will be seen if the drawings are carefully studied at this point, that the portion of the elbow shown by the dotted lines is not required. Patterns for a small portion of section *F*, a certain amount of the surface of section *H*, and the entire surface of section *K*, will be required. The pattern for section *H* is shown in *E* and is developed in the following manner: Draw any vertical line, as *a'b'* in the pattern, upon which place the stretchout of profile *G* in the elevation. Thru these points, at right angles to *a'b'*, draw measuring lines, as shown. Next, from point *f* in section *H* in the elevation, draw the line *fd* at right angles to *hs*. Now, measuring from the line *fd*, take the various distances to the points on the miter lines *hf* and *t–m–4*, and place them on similarly numbered measuring lines in pattern *E*, measuring in each case on either side of the line *a'b'*, as shown. A line traced thru these points will complete the required pattern. The curved outline of pattern *E*, shown

Problem 112. Round Ventilator with Square Base.

by *4'–h'–4'*, will also serve as the miter cut for section *K*.

When developing the pattern for section *F*, shown at *P*, it will be seen that it only requires the stretchout of profile *G* from point *m* to *4'* to *m'*, and that point *4'* represents the position of the central and longest measuring line of the pattern. Place the stretchout upon the line *m'm*, and develop the pattern in the manner described for pattern *E*, measuring from the line *tn* in section *F* to the various points on the miter line *mt* and the curved line of intersection *m–4*.

The pattern for the opening in the straight pipe is laid off at *T*; the divisions from *1* to *7* on the horizontal stretchout line are equal to similarly numbered divisions on the outline of the larger pipe in plan. From the points on the stretchout line, erect vertical lines, which are intersected by horizontal lines drawn from similarly numbered points on the curved miter line in the elevation, resulting in the pattern shape, shown in *T*.

Problem 112. Round Ventilator with Square Base

This problem is presented to show the method employed in developing the patterns for a quadrangular pyramid intersected by a round pipe.

In Figure 151 is shown the elevation and plan view of a round ventilator with a base in the form of a square pyramid that fits over a double-pitched roof. The drawing is shown one-third full size. The various parts of the ventilator are indicated by

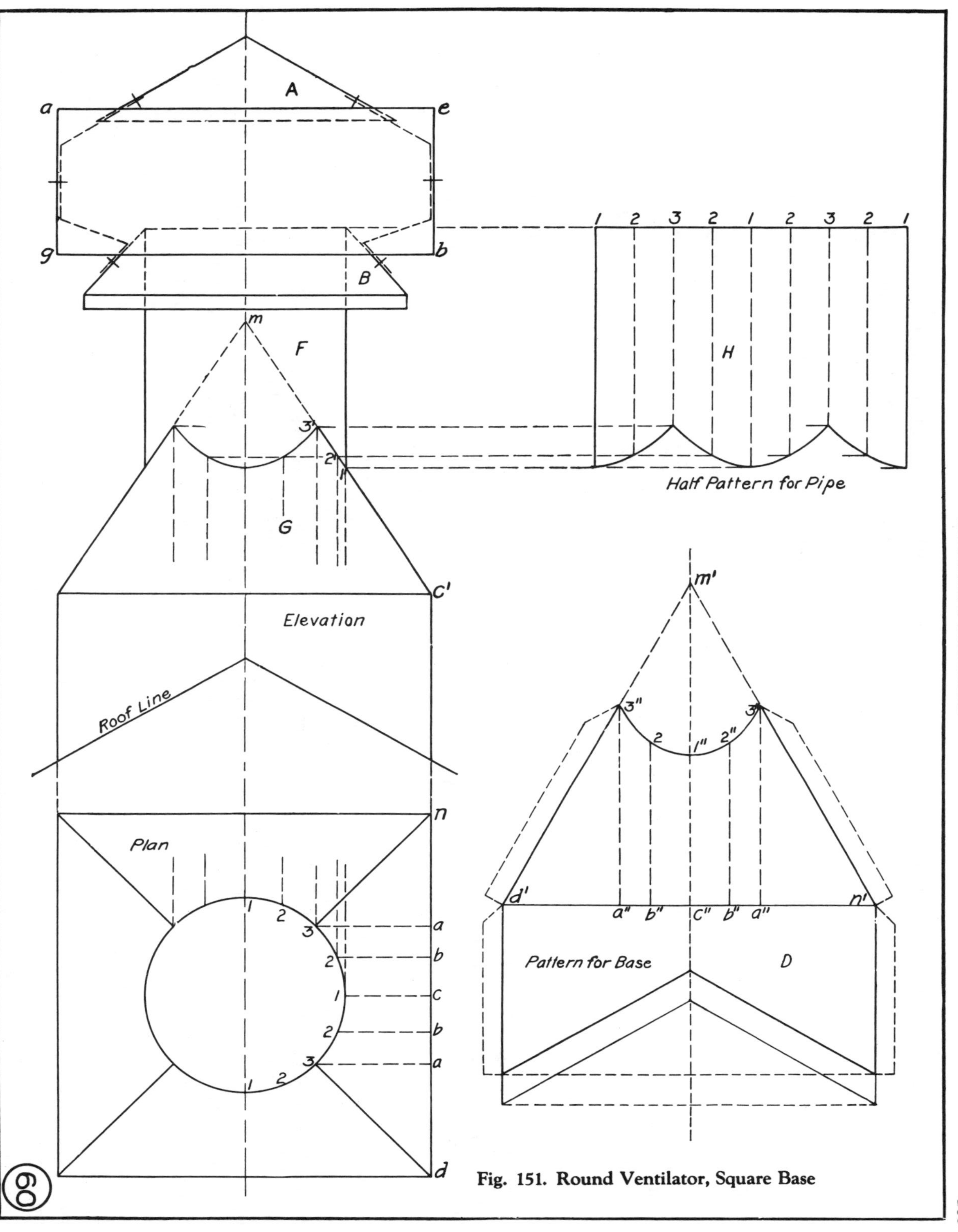

Fig. 151. Round Ventilator, Square Base

the letters, as shown. The upper portion of the ventilator is round in form and consists of the flat cone top, shown in A; the flaring flange or fustrum of a cone, shown in B, which is connected to the upper edge of the round pipe F. A circular band or windshield, shown by $ageb$, is connected to the lower flange B and the top section A by means of a band iron brace, shown by dotted lines in the elevation.

The patterns for sections A and B are developed by the radial method. Since these patterns are constructed in a manner similar to those of preceding problems, definite instructions are omitted.

The principal feature of the problem consists in, first, finding the true line of intersection between the round pipe F and the square pyramid G. After this line has been found, the pattern for the round pipe is obtained in the usual way by the parallel-line method, as shown by half-pattern H. Draw the plan and elevation, and extend the sides of the pyramid in the elevation to the vertex at m. Next, in the plan, divide the circle into a convenient number of equal parts, as shown. These divisions should be so arranged that points will be located on the corners of the pyramid, as shown by 3 in the plan view. From the points 1, 2 and 3 in the plan, draw vertical lines intersecting the side of the pyramid in the elevation at $1'$, $2'$ and $3'$. From these points, draw horizontal lines to intersect vertical lines drawn from similarly numbered points in plan. A line traced thru these points will give the curved miter line in the elevation. This miter line is not necessary in developing the pattern, but is shown here to explain the method of projecting the miter line, no matter what form the pipe, or what pitch the pyramid may have. The pattern for one side of the base with the opening in the top is shown in D.

To obtain the pattern, draw the center line $m'c''$ equal in length to mc' in the elevation. Thru C'', at right angles to the center line, draw the line $d'n'$ equal to dn in the plan view; then connect the points $m'd'$ and n', which will give the full pattern for one side of the pyramid. A separate operation is necessary to find the outline of the opening in the upper portion of the pyramid, where it is intersected by the round pipe.

From the points 1, 2 and 3 of the plan, draw horizontal lines to the base line nd, as shown by 1–c, 2–b and 3–a. The points a, b and c in plan are now located on the line $d'n'$ in the pattern by means of the dividers, as shown by a'', b'' and c''. From these points erect vertical lines, making c''–$1''$ in the pattern equal to c'–$1'$ of the elevation, b''–$2''$ in the pattern equal to C'–$2'$ of the elevation, etc. A line traced thru these points will give the curved outline of the upper edge of the pattern. The pattern for the base is completed by taking a tracing of the straight side of the base in the elevation and placing it in position below the line $d'n'$ in the pattern, as shown. Laps are allowed for seaming and flanging.

Problem 113.   A Round Pipe Intersecting an Elbow Miter

In Figure 152 is shown the method for obtaining the patterns for a five-piece, 90° elbow intersected by a round pipe. It does not matter how many pieces the elbow may contain, or whether the pipe is placed in the center of the elbow or to one side, the principles explained in this problem apply in each case. The drawings in Figure 152 are shown one-third full size. Draw the full elevation of the elbow, as shown by the sections $ABCD$ and E. Then, below the elevation, draw the profile of the elbow and the plan view of the round pipe; also its profile, shown by G'. Next, draw the elevation of the horizontal pipe in its desired position, and draw the half-profile G in size equal to G' in plan.

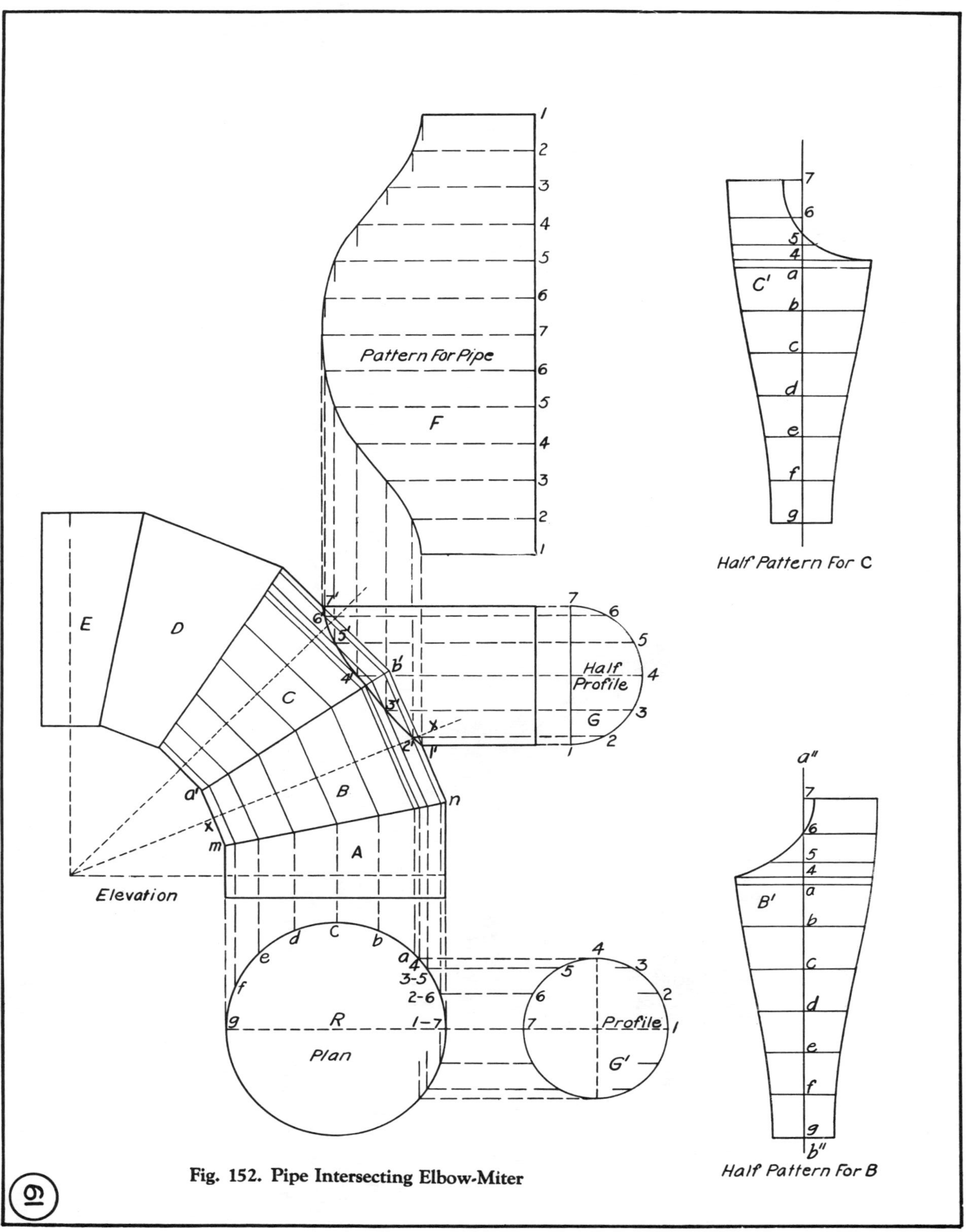

Fig. 152. Pipe Intersecting Elbow-Miter

Divide both of the profiles G and G' into the same number of equal spaces, as shown. From the various points on profile G', draw horizontal lines intersecting the circle R in plan at *4, 3–5 2–6* and *1–7*. From these divisions on R, draw vertical lines intersecting the miter line mn in the elevation, from which points project lines across sections B and C, as shown. Now, from the various points on half-profile G, draw horizontal lines intersecting similarly numbered lines in sections B and C, which have been projected from the plan. A line traced thru these points will give the curved line of intersection between the elbow and the round pipe, shown by *1′, 2′, 3′, 4′*, etc.

The pattern for the horizontal pipe is developed by the parallel-line method in the usual manner, as shown at F in the drawing. A one-half pattern for section B is shown in B', and is developed in the following manner: Draw any vertical line, as $a''b''$ in the pattern, upon which place the divisions g to 7 equal to the divisions g to 7 in profile R in plan. Thru these points draw measuring lines, as shown. Then, measuring from the center line xx in section B, take the various distances to the

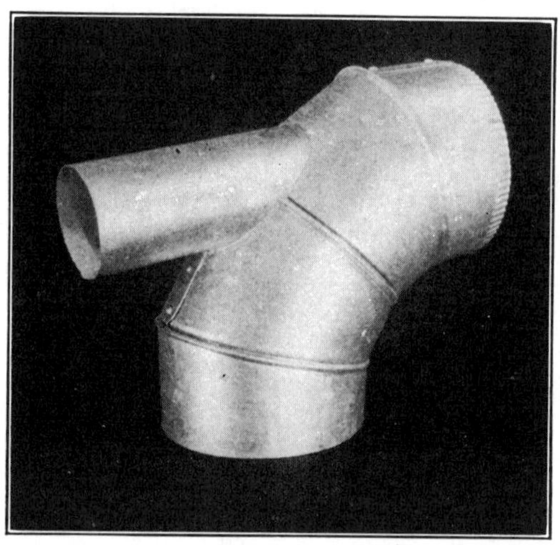

Problem 113. Round Pipe Intersecting an Elbow Miter.

points on the miter lines mn, $a'b'$, and the curved miter line *1′, 2′, 3′*, and place them on similarly numbered measuring lines in pattern B', measuring in each case on each side of the line $a''b''$, as shown. A line traced thru these points will complete the required pattern. The pattern for section C is developed in precisely the same manner, as shown at C'.

Problem 114. Two-Branch Duct Fitting, Square to Rectangular

We have chosen for treatment in this problem the construction of a two-way elbow fitting from square to rectangular, which deals with a special case of square and rectangular duct construction occasionally encountered in the erection of blast systems for heating and ventilating. The large end is square in shape, and each branch is made in the form of a rectangular elbow, the heel and throat being the quadrant of a circle. In Figure 153 is shown the elevation, plan and patterns of the fitting, which are drawn one-fourth full size, and we have for convenience considered the square end of the larger pipe as the base or bottom of the fitting.

Let $FGCEab$ in the elevation represent the front and side views of the two branches of the two-way elbow. The profile of the rectangular end of one branch is shown by the shaded outline $FGHR$. A side view of the other branch is shown by the circular outlines aC and bE, which are described from the centers m and n. The width of the square end of the fitting is shown by the base line CE.

After drawing the elevation, divide the quarter circles *1–11* and *13–20* into equal spaces, as shown. The next step is to draw the plan view directly above the elevation, which shows the position of the two branches; also the miter or joint-line AB. The profile of the square end of the fitting is shown by $ADBO$ in

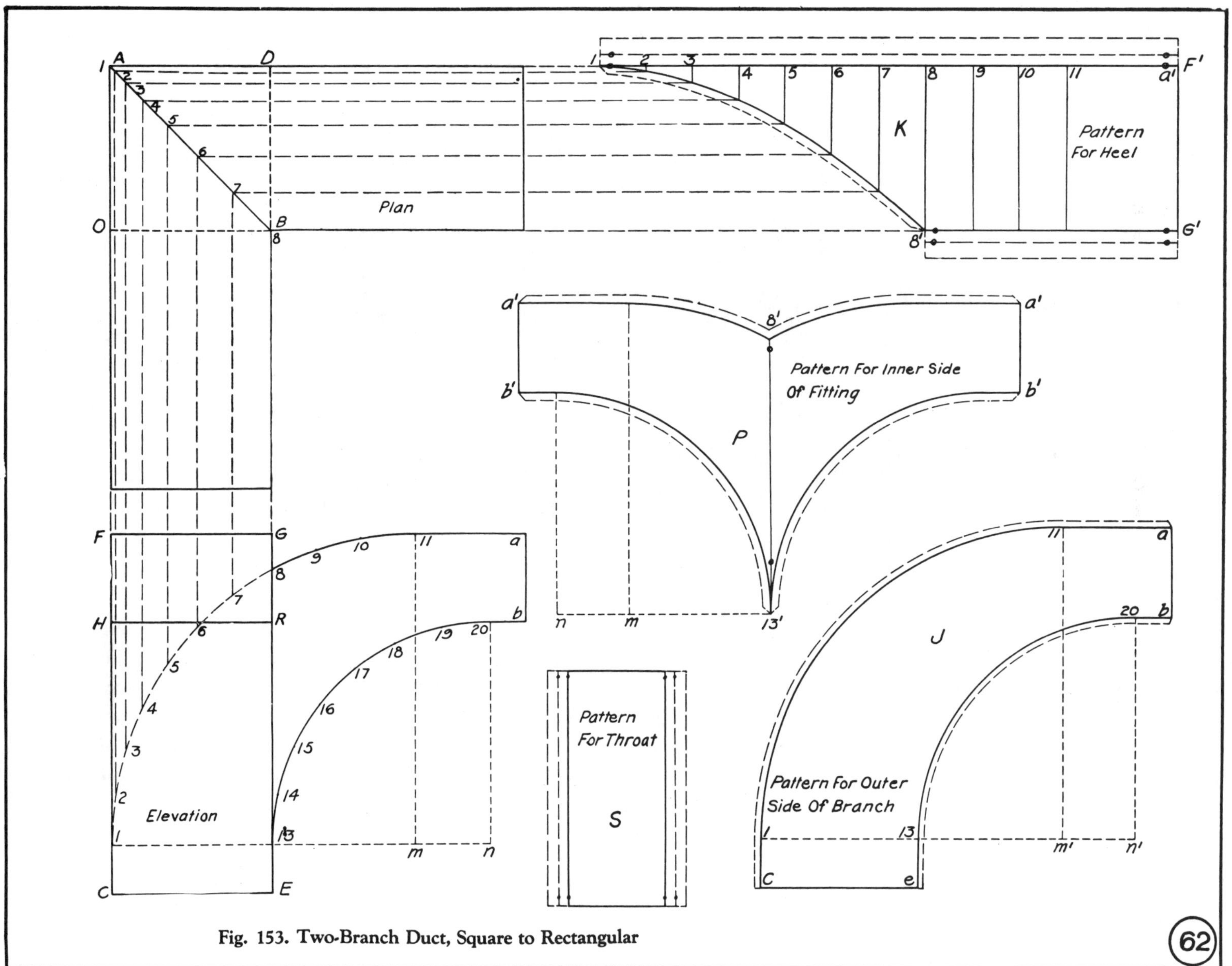

Fig. 153. Two-Branch Duct, Square to Rectangular

62

plan. The pattern for the outer side of each branch is shown in *J*. This pattern is simply a duplicate of the side view of one arm of the fitting, shown by *abCE* in the elevation; the arcs *1–11* and *13–20* are struck from the centers *m′* and *n′*, as shown. Laps are added for seaming, as indicated by the dotted lines. The pattern for the inner side of both branches, which is constructed in one piece, is shown in *P*. The outline of one-half of this pattern, shown by *a′–b′–8′–13′*, is a duplicate of a portion of the side view of one branch, shown in the elevation by *a–b–8–13*, to which laps have been added for seaming.

The pattern of the heel of each branch is shown in *K*. It is developed by the parallel-line method as follows: From the points *1* to *8* in the elevation, draw vertical lines, which intersect the miter line *AB* in plan. Now, draw the horizontal line *1–F* in pattern *K*, and upon this line place the stretchout of the upper side of heel of the branch, the divisions *1* to *11* to *F′* being equal to similarly numbered divisions from *1* to *a* in the elevation. Thru these

points, draw the usual measuring lines, which are intersected by horizontal lines drawn from corresponding points on the miter line *AB* in plan, extending point *8′* to *G′*, as shown. A line traced thru the intersections from *1* to *8′* will give the curved miter cut in the pattern for forming the two upper sides of the branches at an angle of 90°.

The pattern for the throat, shown in *S*, is rectangular in form. It can be laid out directly upon the metal, the length being equal to the distance from *E* to *b* in the elevation; the width is equal to the long side of the rectangular ends of the fitting, shown by *FGHR* in the elevation. Edges are added to the patterns for seaming, as indicated by the dotted lines. Single edges have been allowed on patterns *J* and *P*, and double edges on patterns *K* and *S*, for joining the various pieces by means of a "Pittsburgh seam." This is a quick method of seaming the corners of duct elbows without double seaming or riveting. It is easily constructed and makes a tight, rigid joint.

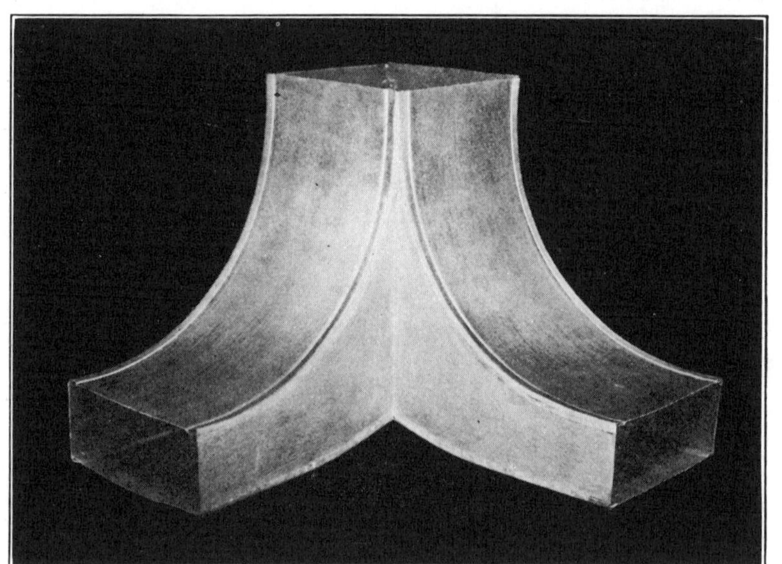

Problem 114.   Two-Branch Duct Fitting, Square to Rectangular.

INDEX